Introduction to Engineering Research

Synthesis Lectures on Engineering, Science, and Technology

Each book in the series is written by a well known expert in the field. Most titles cover subjects such as professional development, education, and study skills, as well as basic introductory undergraduate material and other topics appropriate for a broader and less technical audience. In addition, the series includes several titles written on very specific topics not covered elsewhere in the Synthesis Digital Library.

Introduction to Engineering Research
Wendy C. Crone
2020

Theory of Electromagnetic Beams
John Lekner
2020

The Search for the Absolute: How Magic Became Science
Jeffrey H. Williams
2020

The Big Picture: The Universe in Five S.T.E.P.S.
John Beaver
2020

Relativistic Classical Mechanics and Electrodynamics
Martin Land and Lawrence P. Horwitz
2019

Generating Functions in Engineering and the Applied Sciences
Rajan Chattamvelli and Ramalingam Shanmugam
2019

Essentials of Applied Mathematics for Scientists and Engineers
Robert G. Watts
2007

Project Management for Engineering Design
Charles Lessard and Joseph Lessard
2007

Relativistic Flight Mechanics and Space Travel
Richard F. Tinder
2006

Introduction to Engineering Research
Wendy C. Crone

ISBN: 978-3-031-00955-6   paperback
ISBN: 978-3-031-02083-4   ebook
ISBN: 978-3-031-00155-0   hardcover

DOI 10.1007/978-3-031-02083-4

A Publication in the Springer series
SYNTHESIS LECTURES ON ENGINEERING, SCIENCE, AND TECHNOLOGY

Lecture #38
Series ISSN
Print 2690-0300   Electronic 2690-0327

Introduction to Engineering Research

Wendy C. Crone
University of Wisconsin–Madison

SYNTHESIS LECTURES ON ENGINEERING, SCIENCE, AND TECHNOLOGY #38

ABSTRACT

Undergraduate and first-year graduate students engaging in engineering research need more than technical skills and tools to be successful. From finding a research position and funding, to getting the mentoring needed to be successful while conducting research responsibly, to learning how to do the other aspects of research associated with project management and communication, this book provides novice researchers with the guidance they need to begin developing mastery. Awareness and deeper understanding of the broader context of research reduces barriers to success, increases capacity to contribute to a research team, and enhances ability to work both independently and collaboratively. Being prepared for what's to come and knowing the questions to ask along the way allows those entering researcher to become more comfortable engaging with not only the research itself but also their colleagues and mentors.

KEYWORDS

engineering research, technical communications, research ethics, project management, mentoring

To my family.

Contents

Foreword

You may be dipping your toe into engineering research as an undergraduate or you may have decided that a graduate degree in engineering is the right path to pursue. In either case, there are a number of things that you can learn up front that will make your research experience a positive one and will give you more time and capacity to be the most creative and innovative person that you can be. Engineering research is a very different endeavor than the traditional coursework that you have taken up to this point in your academic career. Students often learn about the broader context of engineering research and the ancillary skills needed to be a successful researcher as they stumble across the need for them. However, that unstructured process is wasteful and takes away from opportunities for discovery and innovation. This book provides guidance and resources on topics ranging from reading journal articles and responsible conduct of research to project management and technical communication. It will serve as a supplement to your interactions with research mentors, advisors, and peers as you engage in engineering research.

> **Student Perspective**
> Students who have recently begun engaging in research have fresh and insightful viewpoints on both the context and process of research that is best expressed through their own voices. Throughout this book you will find perspectives from students who are reflecting on their experiences conducting research projects. These insights and comments are intended to give you a review on research from a different lens.

Preface

Both research as an undergraduate and the transition into research as a first-year graduate student is unlike most of the coursework and school experiences that one has had prior to entering into such an undertaking. Although we carry our technical expertise with us, there are often gaps in knowledge. Additionally, the research enterprise itself is foreign. Without the proper guidance and support, many students flounder and struggle to set themselves on a successful course. This seems wasteful of people's time, disheartening to the individuals involved, and ultimately adds to the attrition seen in graduate programs.

Several years ago, I co-authored an article on topics important to the broader context of engineering research based on an undergraduate course in engineering research developed at the University of Wisconsin–Madison.[1] Additionally, summer undergraduate research experiences at campuses and national laboratories have developed accompanying workshops,[2,3,4] courses,[5,6] and even "boot-camp" experiences[7] that help students to find and understand the scientific literature, appreciate the societal impact of engineering research responsible conduct of research, communicating research findings, research careers, and the graduate school application process. This broader training outside the of the specific research experience has been long advocated by the Council on Undergraduate Research as critical to "socializ[ing] students in the research laboratory culture.[8]"

The semester-long *Introduction to Engineering Research* course developed for the Engineering Physics undergraduate degree program and taught at University of Wisconsin–Madison ad-

[1]Cadwell, K., Crone, W., 2008. Training undergraduates in the broader context of the research enterprise, *ASEE Annual Conference and Exposition, Conference Proceedings*, 1364, 1–9.

[2]The Undergraduate Research Center for Sciences, Engineering and Mathematics and the Center for Academic and Research Excellence, University of California at Los Angeles, http://college.ucla.edu/urc-care/. Accessed January 2008.

[3]Wilson, R., Cramer, A., and Smith, J. L., 2004. Research is another word for education, from *Reinvigorating the Undergraduate Experience: Successful Models Supported by NSF's AIRE/RAIRE Program*, L. R. Kauffman and J. E. Stocks, Eds., Council on Undergraduate Research, Washington, DC.

[4]The University of Washington Undergraduate Research Program. http://www.washington.edu/research/urp/, accessed January 2008.

[5]The University of Virginia Department of Science, Technology, and Society Undergraduate Thesis Project, http://www.sts.virginia.edu/stshome/tiki-index.php?page=Undergraduate+ Thesis accessed January 2008.

[6]Katkin, W., 2004. The integration of research and education: A case study of reinventing undergraduate education at a research university, from *Reinvigorating the Undergraduate Experience: Successful Models Supported by NSF's AIRE/RAIRE Program*, L. R. Kauffman and J. E. Stocks, Eds., Council on Undergraduate Research, Washington, DC: 2004.

[7]Bahr, D. F. and Findley, K. O., 2007. An intensive 'camp' format to provide undergraduate research experiences to first year students. *Materials Research Society 2007 Fall Meeting: Session W4: Implementing New Course Materials and Strategies*, November 28.

[8]Merkel, C. A. and Baker, S. M., *How to Mentor Undergraduate Researchers*, Council on Undergraduate Research, Washington, DC, 2002.

dressed the topic above as well as the importance of diversity in research, research collaboration, safety, and intellectual property. This course was later adapted and implemented at Washington State University and University of Central Florida in a National Science Foundation funded effort. The evaluation of the implementations on their campuses showed that "there was a measurable increase in the understanding of undergraduate research in the students at all institutions.[9]" The subsequent work performed showed that the mode of delivery did not influence the student outcomes. "Similar gains in conceptual awareness between each course format and at each institution" were shown with a one-week faculty-led boot camp, a three-day peer mentor-led course, and a semester-long faculty-led course.[10] Thus, I believe that the usage of the content provided in this book can be successfully adapted to a number of different delivery modes.

I wholeheartedly agree with the assessment of Schneider et al. that "By introducing students to the nuances of the research environment, we believe that preresearch courses reduce barriers to involvement and provide confidence and knowledge for all students who participate.[11]" In our evaluations of the Engineering Physics degree program at the UW–Madison, upon which this book is based, the students who completed the program rated their research confidence and skill levels highly. The majority of students felt that they were able to make contributions to a research team, explain their research topic to other engineers as well as non-engineers, document their research, provide their peers with constructive feedback on their research projects, and identify research misconduct issues. They also reported that they gained skills in conducting a literature search, understanding journal papers, conducting a research project, working both independently and collaboratively, utilizing scientific method, dealing with setbacks, giving and receiving feedback, presenting information, and articulating questions.

These topics are also highly relevant to the first-year graduate student. Even if a student has had a prior undergraduate research experience, revisiting topics can lead to deeper understanding and further skill development. My goal is that students using this book, either independently or while engaged in a research professional development program/course, will be able to gain the skills they need to be successful and achieve a high level of confidence in their research capabilities.

Wendy C. Crone
February 2020

---

[9]Burkett, S. L., Lusth, J. C., Bahr, D., Pressley, S., and Schneider, K., 2013. Three training programs for preparing undergraduates to conduct research. *Proc. American Society for Engineering Education Annual Conference*, Atlanta, GA.

[10]Schneider, K. R., Bahr, D., Burkett, S., Lusth, J. C., Pressley, S., and VanBennekom, N., 2016. Jump starting research: Preresearch STEM programs. *Journal of College Science Teaching*, 45(5), p. 13.

[11]Schneider, K. R., Bahr, D., Burkett, S., Lusth, J. C., Pressley, S., and VanBennekom, N., 2016. Jump starting research: Preresearch STEM programs. *Journal of College Science Teaching*, 45(5), p. 13.

Acknowledgments

This book is based on my experiences as a research mentor, graduate advisor, instructor in the College of Engineering, and an administrator in the Graduate School of the University of Wisconsin–Madison. I am grateful to all of the undergraduate and graduate research assistants who worked with me over the years, not only for their research contributions, but also for how they helped me to develop and learn as a mentor. Although I have taught the course "Introduction to Engineering Research" for more semesters than I can count, it would not have been as successful without the help of a number of key individuals over the years. I would like to thank Professors Greg Moses, Jake Blanchard, and Carl Sovinec as well as other colleagues at the University of Wisconsin–Madison for their collaboration and shared vision in developing the Engineering Physics degree program and the research sequence upon which this book is based.

I also appreciate the opportunities I had to interact with students in the Engineering Physics undergraduate program and especially for their phenomenal engagement, performance, and feedback. I am especially grateful to former undergraduate and graduate students whose perspectives, insights, and comments are included in the Student Perspectives. These are included in the book with permission from Grant Bodner, Christopher Coaty, Aidan Combs, Brian Cornille, David Czajkowski, Chelsea D'Angelo, Tom Dobbins, Chris Everson, Thomas E. Gage, Brad Gundlach, Cale Kasten, Matt Klebenow, Brian Kupczyk, Geoff McConohy, Hugh Ni, Blair Seidlitz, Dan Segal, and Vladimir Zhdankin. I would also like to thank my father, Richard Crone, and husband, Alan Carroll, for proofreading drafts, and my editor, Paul Petralia, for both his patience and nudging to help me get this book completed.

Dr. Katie Cadwell, who was a postdoctoral research associate with the University of Wisconsin–Madison Materials Research Science and Engineering Center (MRSEC) and is now a Professor at Syracuse University, helped to collect valuable learning resources in an earlier expansion of the course. She also helped to make aspects of it accessible to students outside University of Wisconsin–Madison, and worked with Prof. Naomi Chesler and myself on a related project connected to the undergraduate engineering design experience. I appreciate the funding support received from the National Science Foundation through the MRSEC (#DMR-0079983 and #DMR-0520527) and the University of Wisconsin–Madison College of Engineering 2010 grant for Transforming Undergraduate Education in the College of Engineering. Any opinions, findings, and conclusions or recommendations expressed in this material are those of the author and do not necessarily reflect the views of the National Science Foundation nor the University of Wisconsin–Madison.

I had the pleasure of serving in several different administrative roles in the Graduate School at the University of Wisconsin–Madison for five years. These roles included Associate Dean for Graduate Education and Interim Dean, where I provided leadership for all aspects of the graduate student experience, including admissions, academic services, academic analysis, funding, professional development, and diversity. At the time, the University of Wisconsin–Madison Graduate School had a diverse graduate student cohort of ~9,000 in over 140 Master's and 100 doctoral fields across the University. I learned an immense amount from my colleagues in the Graduate School and my faculty and staff colleagues across the University who devote time and energy to graduate education. These experiences and interactions also allowed me to see graduate education from a broader perspective beyond that of the graduate programs in the College of Engineering where I have served as a graduate advisor and research mentor for over 20 years. This book draws from this range of experiences to provide the best guidance and advice I can give to those entering engineering research at the undergraduate or graduate level.

Wendy C. Crone
February 2020

Credits

Table 3.1   Adapted with permission from C. Eugene Allen, Emeritus Dean and Distinguished Teaching Professor, and Former Associate Vice President for International Programs, Vice President and Provost, University of Minnesota, Minneapolis, MN.

Sec. 3.7   Strategies for recognizing and overcoming bias adapted with permission from Molly Carnes, Eve Fine, Manuela Romero, and Jennifer Sheridan. "Breaking the Bias Habit." *Women in Science and Engineering Leadership Institute* (WISELI), University of Wisconsin–Madison, https://wiseli.wisc.edu.

Figures 4.1–4.4   Reproduced from Gall, K., Dunn, M. L., Liu, Y., Labossiere, P., Sehitoglu, H., and Chumlyakov, Y. I. (2002). Micro and macro deformation of single crystal NiTi. *Journal of Engineering Materials and Technology*, 124(2):238–245, with the permission of ASME.

Reproduced from Maboudian, R. and Howe, R. T. (1997). Critical review: Adhesion in surface micromechanical structures. *Journal of Vacuum Science and Technology B: Microelectronics and Nanometer Structures Processing, Measurement, and Phenomena*, 15(1):1–20, with the permission of the American Vacuum Society.

Questions on page 85   From *The Thinker's Guide to Engineering Reasoning: Based on Critical Thinking Concepts and Tools*, 2nd ed., ("the work") Richard Paul © 2013. Used by permission of Rowman & Littlefield Publishing Group. All rights reserved.

Page 97   Courtesy of Springer Nature.

Sec. 5.7.1   D.I.S.O.R.D.E.R. Framework used with permission of Lisa Newton, Professor Emerita of Philosophy, Fairfield University.

Page 129   Reprinted by Permission of the National Society of Professional Engineers (NSPE). www.nspe.org

Page 153   Tips for interacting with the public from *Bringing Nano to the Public: A Collaboration Opportunity for Researchers and Museums* by Wendy C. Crone, 2006. Reprinted with permission of the Nanoscale Informal Science Education Network, Science Museum of Minnesota, St. Paul, MN. www.nspe.org

Figure 7.1   From *Escape from the Ivory Tower* by Nancy Baron. © 2010, by the author. Reproduced by permission of Island Press, Washington, DC.

Assignment 8-1   Laboratory-to-Popular assignment adapted with permission from Caitilyn Allen, Professor, Department of Plant Pathology, University of Wisconsin–Madison.

Sec. 8.4.1   Writing Workshop and "Some Suggestions for Responding to a Colleague's Draft" developed in collaboration with Bradley Hughes, Director of the University of Wisconsin–Madison Writing Center.

Contribution list in Sec. 8.5   From *Responsible Conduct of Research* by A. E. Shamoo and D. B. Resnik. © 2009 Oxford University Press. Used by permssion.

Page 203   Photo by Edna M. Kunkel

CHAPTER 1

Introduction to Engineering Research

1.1   WHO IS THIS BOOK FOR?

The information provided within these chapters is designed for both first-year graduate students and undergraduate students engaging in on-campus or summer research opportunities. For those already in a graduate program, some portions of Chapter 2 will not be relevant. For those just beginning to consider graduate study as a future path, the later chapters will provide you important information for undergraduate research you are currently undertaking as well as some insights on what is ahead of you as you transition into graduate school.

Rather than being an exhaustive resource, this book is meant to supplement your interactions with research mentors, advisors, and peers. There are also other numerous references cited and bibliographies provided that will help you to delve into more detail on particular subjects. You should strive to seek out multiple perspectives on critical topics of importance to you as you move through your engineering research experience.

1.2   HOW RESEARCH IS DIFFERENT

Engineering research is a very different endeavor than the traditional coursework that you have taken up to this point in your academic career. Research is a process of discovery, which means that it has a very open-ended quality as a result. This open-endedness may not be something you are as initially comfortable with depending on your background, but the prior knowledge and the skills that you have developed thus far are still valuable and will help you make a contribution with your research.

Discovery is not done in a vacuum. There is nearly always some prior work in an area or related field that can help us build a foundation from which we can launch our work. The research of today builds upon the findings of yesterday. You may find that you are building on work ranging from 5 months to 50 years ago, so understanding what has come before is an essential part of the process. If your purpose is discovery, then there is no point in rediscovering something that is already known and published. Sometimes, however, as part of the process, you may want or need to replicate the work of others, either as a way to learn a technique or to confirm those results.

Research should also be a mentored experience. You will have many people—your peers, those a bit ahead of you in their studies, staff, and faculty—who you will interact with and rely on for direction, advice, and support. In contrast to the image that many have of research, it is not a solitary activity. In fact, much of the engineering research that is done today occurs in a team environment. These teams are frequently interdisciplinary and may include people from a range of engineering and non-engineering disciplines. Working with people from other disciplines helps us to tackle challenges and open research questions that we might not otherwise be able to make progress on alone. The research group that you work within may be a handful of people or an international collaboration that numbers in the hundreds. Either way, cultivating the relationships within this group and connecting with people related to your research, both on and off campus, will be a critical factor in your success.

The undertaking of research is also something we do with our colleague's and society's trust that we will behave ethically. As individuals within a broader community of researchers, we have the obligation to be responsible and honest. This is required in all aspects of the work, from the design of an experiment to the publication of the results. Our analysis must be conducted with an impartial eye; the results must be presented without manipulation; and, discussion of our research with the broader community of scholars and the public must be done with integrity. With these principles in mind, you will have the best opportunity to create new knowledge, advance understanding in your field, and become a respected member of your discipline.

Ultimately, your goal will be to make what is often referred to as a "unique contribution" to your field. This may seem a daunting task as you enter into research, but as you gain more knowledge about your research area you will soon find that there are a number of things that are not known. You, with the help of your research mentor, will be able to identify an area where you can pursue the creation of new knowledge. It will likely leverage the work of those who have come before you, both in the research group you have joined and in the field as a whole, but you will find a way to make a contribution that is your own. Eventually, you will find that you begin to surpass your research mentor in specific knowledge areas and can begin to think independently about new research endeavors to undertake.

1.2.1   ENGINEERING RESEARCH DEFINED

When we hear the word *research* we often think of it as being synonymous with acquiring new knowledge or even developing some "objective truth." Engineering conjures up images in our mind of applications ranging from computers to bridges. For many, engineering implies improving our way of life or driving technological advancement. When the term "engineering research" comes up, it may be hard to reconcile for some. Is it the creation of new knowledge exclusively? Is it the application of new science to existing applications? Is it the development of new applications? The answer is all of the above and more.

The basic commonality we find in all engineering research is that people are trying to answer questions that have not been asked or answered before, to solve problems that humanity will find useful in some way. We do this through a process of inquiry that relies on careful exploration using scientific method. The answers we find may be immediately applicable or they may add to a base of knowledge that will only see application at a much later date.

There is a spectrum of research from basic to applied. In many cases the same type of basic research might be found in both science and engineering departments and collaborations across these disciplines are common in such circumstances. In a report from the National Academy of Engineering, "Basic research in engineering is by definition concerned with the discovery and systematic conceptual structuring of knowledge.[1]" In contrast to basic research, applied research is much more closely tied to an immediate need and may even be conducted jointly or under a research contract with a company. Across this broad spectrum, an engineering research project might be motivated by some esoteric curiosity tied to the long-term needs of humanity or by an immediate need in a particular community. Regardless of the origins of the research question, the tools we use to answer them, and the time frame in which the results will be applied, these are all a part of the spectrum of engineering research that you will find happening on a day to day basis in universities, national laboratories, and industry.

ASSIGNMENT 1-1:
INDIVIDUAL ASSIGNMENT – ENGINEERING RESEARCH DEFINED

Talk to at least three individuals spanning the spectrum of experience with research related to your general field of interest (e.g., undergraduate student researcher, graduate student researcher, postdoctoral researcher, academic staff researcher/scientist faculty member). Ask these individuals to discuss the topic of "engineering research" with you. What makes a good research question? How do they approach conducting their research? What do they find interesting/exciting about research? Write a 500-word summary of what you have heard that includes both the similarities and differences between the answers obtained from your discussions.

1.3   ENGINEERING RESEARCH CAREERS

There are a wide range of research careers available to people with an advanced degree in engineering. These careers occur most prevalently in industry, government, and academic sectors. Even within one of these sectors the types of jobs that involve research can vary dramatically.

One way to explore the range of options for engineering research careers is to take a look at current job postings in your area of study. Looking at the position descriptions can give you an idea of the work activities and job responsibilities. It will also give you an idea of qualifications and prior experience expected. Some positions may require a minimum education level of a

---

[1]National Academy of Engineering. Committee on Forces Shaping the U.S. Academic Engineering Research Enterprise, 1995. Forces Shaping the U.S. Academic Engineering Research Enterprise. National Academy Press.

Master's degree (M.S.), a doctor of philosophy (Ph.D.), and/or a number of years of professional experience. Finding the kinds of positions that might interest you in the future will provide you with the components of a roadmap for the preparation you will want to pursue.

Although you will likely be most familiar with a traditional faculty position in academia from your experience as a student, the range of research-related careers within the confines of academia is quite broad. Within the faculty ranks alone, the emphasis on research varies between positions depending on the type of institution. A four-year college, for instance, might stress engagement with undergraduate research but have lower levels of expectation on research productivity and a larger amount of time committed to teaching. The research and teaching expectations at research-intensive institutions will vary, but they usually stress research with graduate students, and have a higher level of expectation for obtaining grant funding and producing publications. At larger research-intensive academic institutions there are also a number of non-tenured research positions to be aware of. These often carry titles like instrumentation specialist, scientist, and research professor.

Graduate education both at the M.S. and Ph.D. levels is valuable for people interested in a variety of career paths. Ph.D. recipients don't just end up in academia, but are also sought after by industry and government for their expertise and ability to be innovators and thought leaders.[2] Research laboratories span a range of institutions from government laboratories, some with defense-related missions (e.g., Sandia National Laboratories, National Renewable Energy Laboratory, and Argonne National Laboratory, U.S. Naval Research Laboratory), to other non-governmental research labs, some of them with connections to or histories with universities (e.g., Southwest Research Institute, Draper Laboratory, MIT Lincoln Laboratory). Many medium to large companies also have a research (or research and development) department, unit, or segment of the organization—a few of these being quite large and well-known research enterprises (e.g., IBM Research, GE Global Research, ExxonMobil's Research and Engineering Technology Center, DuPont Experimental Station). The types of expertise needed and range of jobs available are quite broad as you may imagine.

In some engineering disciplines, there is a growing expectation that a person complete a postdoctoral experience after completing their Ph.D. and before obtaining that first "permanent" position. Postdoctoral research positions they are most prevalent in academic settings, particularly large, research universities, although they are also available in some industry and government sectors. These positions are usually full-time paid jobs. Some fellowship opportunities are also available for postdoctoral research positions, both in academia and national laboratories.

---

[2]Council of Graduate Schools, 2013, Open Doors with a Doctorate.

ASSIGNMENT 1-2:
INDIVIDUAL ASSIGNMENT – INVESTIGATING ENGINEERING
RESEARCH CAREERS

Identify an engineering research job sector that you would like to learn more about: industry, government, or academia. Find three different job advertisements in that job sector using on-line resources such as monster.com, usa.gov, and chronicle.com. Ideally, these job postings should advertise a research position related to your area of study. Compare and contrast the positions. Consider things such as the education and prior experience required, duties and re-sponsibilities that the position would entail, and location of the job. Choose the position that you find most interesting and identify the kinds of things you would need to do in the next 5–10 years to make yourself an ideal candidate for this position.

CHAPTER 2

Finding the Right Research Position for You

2.1   SOCIETAL IMPLICATIONS OF TECHNOLOGY

Engineers help to shape the world and our personal experiences in it. Engineering design and research impacts nearly every aspect of our lives: the indoor plumbing and sanitary systems we take for granted, the transportation vehicles and networks we utilize to move about our communities and the world, the structures we live and work in, our communication and entertainment systems, the power production and distribution networks we rely on, and the medications and devices that keep us healthy, just to name a few. Engineers also have the ability to address the grand challenges that face society and improve the human condition by doing so. These challenges exist throughout the array of human experience, from ensuring a stable food supply and clean drinking water for the world's population, to the further development of artificial intelligence and treatment of neurological disease by reverse engineering the brain.

It is an exciting and important moment in human history for engineers. The world depends on us, both to maintain our current standard of living and to innovate in new and unprecedented ways to bring us into a better future. We have the capability and the responsibility.

ASSIGNMENT 2-1:
INDIVIDUAL ASSIGNMENT – YOUR ENGINEERING GRAND CHALLENGE

Read the National Academy of Engineering's "Grand Challenges for Engineering" list[1] and identify the challenge most closely related to your research interests. Summarize the challenge and describe the ways in which research in your field of study have already impacted this topic and how you imagine future research can make an impact on this challenge topic.

---

[1] National Academy of Engineering, "NAE Grand Challenges for Engineering," http://www.engineeringchallenges.org.

2.2   IDENTIFYING A RESEARCH PROJECT

Sometime students think that in order to engage in research you have to come up with the idea yourself at the very start. This is quite a challenge if you are new to a field and have little prior experience with research. Identifying a research project that you can undertake usually involves a very different process. Experienced researchers are often looking for students to help them with new and ongoing research projects. So, what you are actually seeking is a match between your interests and existing research projects that are available.

> **Student Perspective**
>   "[R]esearch never really 'ends.' What I mean by this is that even when a group gets a paper published on an experiment it doesn't end there. Frequently, the group continues to do research on the same topic using the ideas and results from their last paper. I guess this does make sense to me, but again it was something I never really thought about. In some way, I suppose I assumed that after one project was finished, they would look for something new and exciting. But, once an experiment is completed, there is almost always further research to be done to learn even more about the topic."

Whether you are an undergraduate student or a graduate student, you should enter into a research project that meshes well with your interests. Don't just take on a project for the money or because it is the first one offered to you. Cast your net wide and look for a variety of projects that might fit your interests as well as a research mentor who would be a good match for your personality and needs. After you find the right research project to pursue, your intrinsic interest will motivate you through the difficult parts and ultimately help you to be more successful.

In order to identify research projects and mentors that are a good fit, first identify the areas of engineering that interest you. Explore your options by reading about current research in those areas and talking to people who have experience with ongoing research. Utilize a variety of sources including websites and recently published journal papers. As you begin to identify individual faculty members you might be able to work with, try to engage in face-to-face or email conversations with these potential research mentors. It is easy to be energized by someone's enthusiasm for their work, but don't fixate on the first thing you learn about. Look broadly and determine what options might be available to you. Even if you are entering a summer research opportunity, rather than a new degree program, often there are choices of projects available to you and faculty mentors within the program that you can identify as your top choices.

Some people stumble across the perfect research position immediately, but often students need to make some effort to both identify potential research mentors and find ones who are willing to add you to their research group. Often available research funding can be a barrier. If you are an undergraduate student looking for research experience, you might choose to do this work for credit rather than pay. That option may open additional opportunities that would otherwise not be possible. Graduate students frequently have the challenge of finding a good match between their interests and the funding available for a research assistantship. If you have obtained a fellowship, this becomes less of an issue, but most students will need to find support either as a research assistant or a teaching assistant.

Consider these strategies if you are having difficulty obtaining a research position.

- Cast a wide net so that you don't limit your options too severely up front.

- Be as flexible with your research interests as is reasonable.

- Consult with faculty you have taken classes from; ask about openings they may know of or colleagues they would recommend.

- Seek out new faculty (e.g., assistant professors) who may be looking to grow their research group.

- Identify research centers or facilitates that may have positions available.

After you have explored what is available to you, some introspection will be called for. If you find that you have developed a keen interest that is not represented at your institution, you may have to consider making a change. As an undergraduate, you can consider looking for summer research opportunities elsewhere, transferring to another institution, and/or pursuing your later graduate studies at an institution with a better fit with your interests. As a graduate student, hopefully you will have taken on this exploration while looking for the right graduate school for you, but, if you find yourself at an institution where your interests are not represented, you have to make some decisions. Stay or go elsewhere? Some programs allow for a "coursework only" Master's degree that you can finish up more quickly so that you can move on to another institution sooner. If you can find a research project peripherally related to your interests, you might want to consider pursuing this for your Master's degree research and then make a change when you begin your Ph.D. or first industry position. This is not as unusual as you might think. I have known many students who have made a significant change after their Bachelor's or Master's degree. Their prior experience is not a waste, they will be able carry their skills and knowledge forward and may be able to use them in unanticipated ways.

Some students find themselves paralyzed at having to choose which research project they will take on. If you find research areas at your institution that excite you—which is often the case—you may find that you have more options that you expected. The important thing to remember is that it does not have to be a decision you are married to forever. Although it is likely that your research career will be related to the general area of study you are currently pursuing, it is also likely that your research career will be long and varied. The research I did as an undergraduate was in the same basic field as my graduate work, but not thoroughly connected to it. Also, the specific research I did for my master's degree was different from my Ph.D. (and different from what I do now as a faculty member). You can choose to stay in the same area or you can use the skills you have learned in related areas. You will find that much of what you gain in both your coursework and research experience is transferable and can be used in other areas of engineering application.

There are often opportunities to move around and try new things as you progress in your studies and career. Technology also moves quickly, so even if you begin your career in a particular specialty area, it is likely that you will have to learn and expand your expertise over time. Outside of academia, change is even more common—switching between companies or organizations, working in different positions—and often require different competencies and your own personal career management.[2] Most researchers, even faculty researchers, change their research focus over the course of their careers even if they stay at the same institution.

Who your research mentor will be is as important as the topic of your research project. Research mentor fit is often overlooked, but as Megan Poorman, GradHacker blogger, points out: "Choose your mentor wisely: this is the biggest factor in your job satisfaction and degree progress. Your advisor sets the tone for the lab and can either help or hinder your professional development and your research progress. Find someone with whom you can communicate and who will be on your side, looking out for your best interests. I would choose the mentor over the research project. Obviously, you should be excited about the research, but projects change and morph over time, your mentor likely will not. Choose wisely.[3]"

> **A Research Mentor Who Wants You to Succeed**
>
> Some of the proudest moments in my professional life have been because of the success of my students, either currenter or former. When they give a fantastic research presentation, earn a prestigious award, win a fellowship, get their dream job, or achieve the promotion that they were seeking, I feel great pride. I hope that in some way I have helped them to make these successes for themselves. Although I have been described variously as sympathetic, supportive, and demanding as a research mentor, these are consistent descriptions, given that my goal is to figure out the needs of each of my students and help them to be their best and achieve their goals. But when it comes right down to it, each individual is their own person figuring out who they are and who they want to be. You need to find the right research mentor for you who will help you be your best and work towards your goals.

Consider some of the following questions when you are interacting with potential research mentors.

---

[2]Seibert, S. E., Kraimer, M. L., and Crant, J. M., 2001. What do proactive people do? A longitudinal model linking proactive personality and career success. *Personnel Psychology*, 54(4), 845–874.

[3]Poorman, M., 2019. GradHacker, "Hacking Grad School," Inside Higher Ed. https://www.insidehighered.com/blogs/gradhacker/hacking-grad-school.

- How much time and attention do you need and does it match with the potential research mentor's availability?

- Does this individual provide the following to their research students:
  - constructive feedback?
  - assistance in setting realistic goals?
  - feedback about expectations?
  - information about funding opportunities?
  - professional development opportunities and connections?
  - aid with the job search?

- Do you need someone who will be encouraging and nurturing or are you more comfortable with a higher level of autonomy and independence?

- Do more experienced students and the graduates from the research mentor's group develop professional independence and transition to the status of junior colleague?

- If you are interested in a particular career outcome after your degree, will this mentor be able to support this interest and help to launch you on this trajectory?

Even when you find the "perfect fit," it is important to realize that you will need to develop other mentors beyond your primary research mentor throughout your research experience. The pool of possible mentors is large and includes other faculty members, research staff members, postdoctoral researchers, as well as other students at both the graduate and undergraduate levels.

Student Perspective

"Other goals that might help me to become an independent researcher include making sure to seek the advice of the experienced researchers and research mentors I may work with in the future, and staying honest with myself about who I am and what I want. Taking guidance from mentors and forming close relationships with them seems to me to be one of the main ways people find their place in the research world. Mentors know how the research world works and can give good advice to young researchers on what steps to take to get where they ultimately want to be. This is where the second goal is important. I want to remain conscientious about where my path leads me and to make sure at all times that I am not being funneled into an area or profession that will be unfulfilling. I don't want to look back in my middle ages and wonder what happened."

ASSIGNMENT 2-2:
GROUP ACTIVITY – RESEARCH INTERVIEWS WITH OTHER STUDENTS

Overview: This activity will give you the opportunity to find out about the research that others are interested in and express your own interests about research. The objectives for this activity depend on your prior experience.

1. For students with research experience: you will have the opportunity to practice your communication skills in the context of the research you are conducting and reflect on the progress you have made as a researcher.

2. For students inexperienced with research: the interviews will give you the opportunity to learn more about the kinds of research being undertaken on your campus. The in-class interview activity should also help to increase your comfort level when talking to potential faculty research mentors outside of class.

Preparation: be sure take time to think about the following in preparation for the interviews.

For students inexperienced with research:

- Brainstorm questions that you might ask. (Note: you will be doing more than one interview and you will be conducting interviews with students having different levels of research experience.)

- Some suggested started questions might include the following:

  – What is your area of research?
  – How did you get involved with the research are you are currently working in?
  – Has your research experience been what you expected?
  – Have you run into any stumbling blocks in your research? How did you overcome them?
  – What approaches would you suggest for finding a research project?

- You will need to listen actively and do your best to ask probing follow-up questions based upon hearing the initial response.

For students experienced with research:

- Consider the starter questions posed above and what you feel would be the most valuable information to discuss with a student inexperienced with research.

- Use the strategies discussed in Chapter 7 to organize your thoughts.

ASSIGNMENT 2-3:
INDIVIDUAL ASSIGNMENT – PROFESSIONALISM IN EMAIL INTERACTIONS

There will be many occasions throughout your research career where you will need to initiate contact with someone via email. This is an important opportunity for you to make a good first impression by displaying professionalism in your email communications. It can be a big mistake to approach this initial interaction casually or sloppily.

Write precisely and clearly so that your meaning is understood. It is not appropriate to include emoticons or emoji, but you want to be sure that your message comes across with the right tone. Don't use humor or sarcasm. Check your spelling and grammar. Err on the side of formality. The person reading the email will make a lot of assumptions about you based on the limited information that the email contains. You want to ensure these assumptions are as positive as possible.

The message should begin with a salutation. "Dear Prof. Smith" is appropriate, "Hey" is not. State your request up front. Tell the person who you are and why you are making this request. Indicate how you would like to follow up (e.g., "I would appreciate it if we could set up a meeting. I am available….", or "Thank you in advance for your reply.") Your message should include enough information to be clear, but not be so long that it will not be read.

At the bottom of your message you should have a signature block with your contact information. Something simple like the following will suffice:

Ima N. Gineer
Undergraduate Student, Engineering Mechanics Program
University of Wisconsin-Madison
Cell: 999-999-9999

Your assignment is to compose an email to a faculty member using the above guidance. This message should do one of the following.

- Request an opportunity to meet and discuss the research being undertaken in their research group.

- Identify a research question related to a published journal article that you have read and request guidance on what additional follow up reading might help you to answer your question.

- Inquire about the availability of a research position.

- Pose a question about an area of content of a course you are currently taking and request their guidance.

- Inquire about an interesting course that you anticipate they will teach in the future.

ASSIGNMENT 2-4:
INDIVIDUAL ASSIGNMENT – CONVERSATIONS WITH POTENTIAL FACULTY RESEARCH MENTORS

STEP 1: Identify five faculty members you will contact about research projects. In addition to identifying their name and research interests, find their contact information, including email address, phone number, and office hours.

STEP 2: Summarize the area of research that each faculty member specializes in. Look for a recent news article, webpage summary, or journal publication to give yourself a bit more background about their work. Note that often faculty research interests change over time although web pages may not be revised frequently, but this information will at least provide you with some relevant background about their research interests.

STEP 3: Draft an email of introduction. Use professional language, including the appropriate salutation (e.g., Dear Prof. Smith). Consider attaching a resume showing your prior work experience—even if your work experience is not research-related, it shows that you can hold a job and perform it reliably. Indicate in your message how you will follow up with contacting them (e.g., I plan to visit your office hours next week so that I can learn more about your current research interests). After using spell check, send your emails.

STEP 4: Follow through with your follow up! Ideally you should talk to the faculty members you have contacted either in person or by phone. Come to the conversation prepared to do the following:

- Describe what you find interesting about the research they have done.

- Discuss your experience and interests.

- Ask about their current research and future research interests.

- Specify what you are hoping for as a result of the conversation.

2.3   UNDERGRADUATE RESEARCH EXPERIENCES

If you want to learn about research, it is a great idea to start early while you are an undergraduate student. There are many advantages to doing research as an undergraduate—you can learn about the process of research to determine if this is something you are interested in doing more of, you can try out a particular research area to see if it is something you would like to pursue further, and you can gain some basic research experiences that will be to your advantage when you apply to graduate programs.

> **Student Perspective**
> "This experience [as an undergraduate researcher] was a valuable one. It taught me a lot about myself and what I really wanted to do and was interested in. It also gave me a great look at one style of lab organization in terms of people and project roles within the group. I was able to work on and realize the importance of networking and general looking out for myself in research."

The types of research positions available for undergraduates on university campuses vary. They range from "bottle washer" positions to those that involve doing an independent research project. Often it is the case that a research position is a combination of different tasks at a variety of levels, from glamorous to tedious. (Someone must wash the glassware, right?) Undergraduates are often hired into research labs to help out with some of the work that might be a little bit more routine, but these are still great research opportunities because it allows you to learn about the work taking place in that research group and gives you the potential to work your way up, and take on more responsibility, as you prove yourself to be capable and dependable. Additionally, undergraduate research is usually a bit lower stress and forgiving of failure.

> **Student Perspective**
> "I think one of the main expectations that the group has for me is that I'm not afraid of failure. By this, I mean that the project I'm working on has never been done the way they are asking me to do it. Because I am not a Ph.D. or Master's student, I am the perfect person to conduct the experiment because I don't have any pressure to produce publishable results and I'll be able to focus more on the research at hand. Although they do have high hopes for the project I'm working on, I won't have the pressure that the typical Ph.D. or Master's student would have. So, I guess my biggest goal for my project is to produce results that the group can do something with. But, also to be optimistic if they don't always turn out as I had hoped."

Many undergraduates find meaningful research experiences on their home campus. There are a variety of different ways to connect in with research, and a variety of ways that you can go about getting compensated beyond the experience you will gain. You can look for jobs that are paid positions. These range from entry-level positions that pay minimum wage to more high-paying positions that use your technical skills. This may begin as a part-time job, where you are assisting with day-to-day needs in a laboratory and grow into a research experience as you develop your skills and show initiative. Or, you may have the opportunity to conduct research for credit. For instance, as an independent study project under the supervision of a faculty member. Some campuses also offer scholarship or fellowship opportunities connected to research. Often these kinds of opportunities will allow you to propose a specific research project with a research mentor and apply for some funding to complete that research activity. If you have a research area(s) in mind that you would like to get experience with, you might be able to find a research group working in that area that would be willing to let you attend group meetings and/or spend time shadowing a graduate student. Your academic advisor will be able to give you information about the options available to you on your campus and how to go about pursuing them.

In all of these cases, you need to be able to devote enough time to do the research to make it worthwhile for both you and your research mentor. I suggest that you need to devote at least 10 hours per week so that you can spend enough time to become competent and productive. That also means putting in time *every* week in order to make research progress. In a paid position you will be paid by the hour. If you are getting course credit the expectation is usually a minimum of 45 hours per semester credit. If you assume a standard length semester and three credits of research, this would be roughly equivalent to 10 hours per week. At the end of the semester you will likely need to produce some kind of document, like a report or poster, which summarizes your research project and the progress that you have made.

Now the question is how to find a research position. The first thing you want to do, before you start sending emails and knocking on doors, is to figure out what kind of research is of most interest to you. Take a look at what kind of research is being conducted on your campus. The websites of faculty members, research groups, and research centers can provide useful information, but keep in mind that the research projects that are actively being conducted may not be represented on the website yet. Although the projects being discussed on the website may not be ongoing, it should still give you a flavor for the type of research being done in that research group.

The next task is to prepare yourself: put together a professional-looking resume. If you don't know where to start, the career services office will likely have helpful information, and possibly even workshops to assist you in creating a resume to highlight your experiences and skills. You also need to be prepared to talk to a faculty member about your research interests, as well as your background and capabilities. You should not just show up and say "Hi, I want a job." You need to be able to articulate your interest in the research area that faculty member is engaged in, and talk about the qualities and capabilities that you could bring to the research group. You may not think you have much to talk about without prior research experience, but you may have qualities like dependability, skills that you have developed through hobbies, and background that you have obtained in courses, that you can speak about.

It is likely, however, that having access to these kinds of opportunities will require some persistence on your part. Research positions for undergraduates on most campuses are relatively rare. If you throw up your hands and give up at the first obstacle, you will be unlikely to find the sort of experience that you are interested in. This is also good preparation for doing research, because doing research will require persistence and the ability to work your way around the roadblocks that appear.

There are a variety of strategies that you could employ and you should begin with the one you feel most comfortable with.

- One way to initiate contact is to send an email of introduction with your resume as an attachment. Even if you don't hear back from the faculty member right away, you can then follow up during the faculty member's office hours.

- You can talk to your academic advisor to find out if they have a suggestion for who may have research openings that you can apply for.

- You may have friends and classmates who are already involved in research. You can talk to them about whether there are openings in their research group, and if they would make an introduction to their research mentor on your behalf.

- You can talk to professors you have taken classes from and have done well in. They might have a research opening, or they might be able to suggest who would.

- Use your network. Talk to people about your interests and what you would like to do. You never know, the person you see on the bus every day, or the person you know from your soccer club, might be just the contact you were looking for!

Expect that it will take more than one contact attempt with a prospective research mentor, as well as more than one potential research mentor on your contact list.

Student Perspective
"I tried once again to reach this professor through email, but realized that I'd have to try for an in-person meeting if I was going to get anywhere. I quickly learned that this professor was extremely busy. I spent a lot of time that semester waiting in the hallway to meet with him. After more than a month of rescheduled or missed meetings, I got an interview and soon began work ... I think the clearest lesson I learned during the process of getting [a research position] was that sometimes you have to be a little impertinent to get noticed."

An alternative or additional way to gain research experience as an undergraduate is to apply for a summer undergraduate research program. These are often called *research experiences for undergraduates*, but they have various titles and are offered by a number of different organizations. For instance, the National Science Foundation (NSF) sponsors numerous research experiences for undergraduate (REU) programs around the country, mainly based at university campuses. Several national laboratories offer summer research opportunities, such as the Sandia National Lab Summer Internships and the NIST Summer Undergraduate Research Fellowship (SURF). There are also a few international opportunities to do research such as the DAAD Research Internships in Science and Engineering (RISE) that sponsors U.S. student to go to Germany for research opportunities. In your conversations with your academic advisor and professors, you can ask about summer opportunities that they might know of at other campuses and institutions.

These summer programs are almost always paid opportunities and there is usually some coverage for living expenses. The quality of the research experience can vary, so you will want to be sure that the ones you are applying to provide authentic experiences in research. Look into the range of different things that might be available to you. If you are persistent about seeking them out, you are likely to find a really great research position.

Regardless of the specifics of the position, and how it is compensated, you should approach it in a professional manner. Once you obtain a research position you need to be responsible with how you conduct yourself and how you take on the work. Ideally, you will also show initiative by thinking creatively and innovatively about the details of the project. As you exhibit these traits within a research setting, you be given more responsibility as time goes on. If your contributions are not noticed, then you need to point them out and ask for more responsibility so that you can show what you are capable of contributing. Research shows that that being proactive is directly linked to career success and satisfaction.[4]

> ### Sometimes You Don't Have to Make a Choice
>
> One of my undergraduate advisees, who is already engaged in an extensive on-campus research experience, is now thinking about the tradeoffs between gaining more research experience in a different area through a summer research program vs. studying abroad. It's a tough choice, but my main piece of advice is that she may not need to choose. It may be possible for all of it to happen, just over a larger time span than she originally imagined.
>
> It is easier, and more common, to study abroad as an undergraduate than as a graduate student. However, it is possible to put off study abroad without giving up the opportunity all together. I spent a semester in Australia as a graduate student, but I had to independently organize it rather than join an orchestrated program with multiple students. There are pluses and minuses to the differences in those experiences, but both will give you an opportunity to immerse yourself in another culture.
>
> Summer research experiences can be a great way to get experience with another area of research and another institution. If there is a specific area of

[4]Seibert, S. E., Kraimer, M. L., and Crant, J. M., 2001. What do proactive people do? A longitudinal model linking proactive personality and career success. *Personnel Psychology*, 54(4), 845–874.

research you are interested in exploring a bit more prior to graduate school, you can look for a program at a different institution that would provide you with an experience related to your interests.

The other consideration may be money. Summer research programs usually pay several thousand dollars and sometimes provide you with a place to live. Study abroad programs are generally something you must fund yourself as an undergraduate student. (There are some exceptions, with a few scholarships that are available, and funding opportunities for graduate students to do research abroad.)

2.4 THE GRADUATE SCHOOL APPLICATION PROCESS

2.4.1 IS GRADUATE SCHOOL RIGHT FOR YOU?

Graduate school is an excellent way to continue your education, deepen your engineering skills, and open yourself to other career opportunities. However, graduate school should not be viewed as simply an extension of your undergraduate studies. In most cases, earning a Master's degree or Ph.D. will take more than a few extra classes. Particularly for the Ph.D., it takes an interest in and serious commitment to research. When considering applying to graduate school, examine your motivations. Because you are not sure what to do next, don't want to venture into the "real world" yet, or think the job market is tough are NOT good reasons to go to graduate school. In fact, these unsuitable motivations will likely show in your graduate school application materials and make it very difficult for you to get accepted.

That being said, I encourage all of my advisees with good GPAs to seriously consider graduate studies. Maybe it is not something they are interested in embarking on right away, but it should be kept in mind in the coming years. Of engineers holding a B.S. degree, 40% go on to get a Master's degree and 9% go on for a Ph.D.[5] Many companies will consider whether or not someone has an advanced degree at hiring and/or promotion. Some companies will even provide funding for courses and/or a graduate degree.

Once you have decided to consider graduate studies, then you need to decide if you want to apply to get a Master's degree (often called terminal Master's) or to a Ph.D. program where you will likely complete a master's degree on your way to your Ph.D. It is alright to not be 100% certain of your goals at the point of application, but you should represent yourself honestly and indicate how strong your desire is to continue on for a Ph.D. In many programs this can be a deciding factor for entry and for funding, so you should try to choose programs to apply to that will be a good fit for your goals.

---

[5]National Academies Press, Understanding the Educational and Career Pathways of Engineers, 2018. https://www. nap.edu/catalog/25284/understanding-the-educational-and-career-pathways-of-engineers.

Going Corporate

After I completed my Master's degree, I began working in industry (which I really enjoyed). As I learned more about the company and the engineering positions available, I realized that I was keenly interested in research and development. My main motivation for returning to graduate school was that the jobs that I was most interested in obtaining in the future required a Ph.D. This was a strong motivation to go back to graduate school. It turned out later on that I found teaching and research to be my passions, so I never did go back for that industry dream job that I had my eye on. Sometimes my path has not been a linear one, but all the experiences I gained along the way have been valuable.

ASSIGNMENT 2-5:
GROUP ACTIVITY – GRADUATE SCHOOL FIT

In small groups, brainstorm about what qualities are important for being successful at graduate-level research. Share these with the class.

In small groups, discuss why you might pursue graduate studies immediately after completing the B.S. degree or wait 2–5 years; advantages and disadvantages of both. Share these with the class.

ASSIGNMENT 2-6:
INDIVIDUAL ASSIGNMENT – GRADUATE SCHOOL APPLICATION EXPERIENCE

Identify a current graduate student in the field of study you are interested in pursuing. Talk to that person about their experience in applying to graduate schools.

2.4.2   THE GRADUATE SCHOOL APPLICATION PACKET

Understanding the main components of the graduate school application packet, well in advance of when you plan to apply, will help you to build the strongest application possible. The main pieces of most application packets will be information from your undergraduate institution, such as your grade point average (GPA) and transcript, your Graduate Record Examination (GRE) scores (if required), and letters of recommendation. You will also need to write one or two essays for the application where you will commonly be asked to describe experiences that make you well suited for this graduate program and your long-term goals related to the pursuit of this advanced degree.

Nearly all graduate school application will require letters of recommendation (usually three, sometimes four). These are important because they are often the best predictor of whether or not an applicant will be successful in a particular graduate program. Some of these letters will be written by faculty members—ideally ones who have interacted with you on a research project, a student organization/team, or as an instructor (ideally in more than one class). It is also relevant to ask a supervisor or manager from a current or prior work experience even if it is not specifically engineering related (they can speak to issues such as reliability and initiative). You may also have some more extensive involvement in a volunteer activity. A letter from someone in authority in that organization might also prove useful.

Help your recommenders write you the best letter possible. Give them plenty of advance notice and a reminder when the deadline is a few weeks away. Provide them with materials to refer to such as your resume and/or your application essay(s). Remind them explicitly how you have interacted previously (e.g., "As you may recall, I took Advanced Mechanics of Materials from you last Fall semester and my team completed a design project on…"). Provide them with a list of items you would like them to address in your letter (e.g., "I am hoping you can speak to the work I did in the lab over the last several years, especially the project where I refurbished the testing equipment and developed new protocols for operation. In addition to working in the lab 15 hours a week, I was also a member of the Marching Band and maintained a 3.5 GPA.").

ASSIGNMENT 2-7:
INDIVIDUAL ASSIGNMENT – GRADUATE SCHOOL APPLICATION REQUIREMENTS

Identify a graduate program that you would like to apply to and determine the deadline for application and the application materials you will be required to submit.

Look for requirements such as minimum GPA and GRE test scores (usually just general, but some programs require a specialty exam). Determine what documents (e.g., transcripts), essay(s), and letters of recommendation will be needed. Read the instructions to determine if there are any specific expectations for what should be addressed in the essay(s), and if your resume should also be included in your application materials.

2.4.3   THE APPLICATION TIMELINE

A common timeline for the graduate school application process.

Summer/Early Fall

- Identify some graduate programs that you are interested in applying to and identify the application requirements and deadlines. Determine whether or not the GRE General Test and GRE Subject Test are required (although the General Test is usually expected, the Subject Test is not common for most engineering graduate programs).

- In preparing for the GRE General Test, I do not generally recommend spending money on a preparation course. Your score will be close to its maximum if you take a few practice exams to familiarize yourself with the test format and the way in which questions are posed. Educational Testing Service (ETS) offers free practice tests and software which you can use to emulate the actual test environment (see `http://www.ets.org/gre`).

- By the end of October you should have taken the GRE (although it is available year-round).

Mid Fall

- Identify and contact people who will provide you with letters of recommendation (see above for ideas about who you should consider asking for a letter).

- Finalize the list of programs you will apply to. Identify faculty members in each of these programs whose research you find interesting and initiate contact with them by email or phone.

- Begin preparing your applications. Look for graduate school application workshops and/or a faculty member who will read over your application materials for you and provide you with feedback.

Late Fall/Early Winter

- Complete your applications and submit them BEFORE the published deadline. Often, the review of applications begins prior to the cutoff deadline and you would like your application to receive the fullest consideration.

- Thank your recommenders for taking the time to write letters of recommendation for your applications. Send a brief note of appreciation—ideally in an "old fashioned" thank-you card, or at the very least via email.

Winter

- Follow up with faculty members in the programs that you have applied to. Contact only those who you are keenly interested in working with, but be persistent in attempting to get through to them. If your email message does not get a response, then make a phone call. Also, consider asking your letter writers if they know any of the individuals you have identified and ask if they would be wiling to write an email of introduction for you.

- Many departments offer a visit weekend for prospective graduate students. Ask the department student services coordinator or faculty you have been in contact with if there would be an opportunity for you to visit the campus and meet with faculty and students. Often, some or all of the travel costs are paid for, but, even if they are not, you should make your best effort to attend.

Late Winter

- Attend prospective graduate student visit weekends that you have been invited to. Meet with faculty and graduate students and gather as much information as possible. It is a two-way interview: you are trying to present yourself in the best possible light and you are trying to determine if this graduate program is a good fit for you. See the list of "Questions to Ask Yourself and Others" below.

Spring

- Consider the offers that you have received. Note that some programs make separate offers for admission and funding, so be certain that you understand the implications of each offer.

- YOU CAN ONLY SAY YES to one. Nearly all universities in the U.S. are members of the Council of Graduate Schools and honor the April 15th resolution.[6] This means that students should not be obligated to respond to an offer prior to April 15th. This gives each student an opportunity to see all offers available to them prior to making a commitment. Additionally, this means that **you can only accept one offer**. A student who accepts an offer has made a commitment and should not accept any other offer without getting a written release.

- Inform your advisor and recommenders of your decision so that they know where you are going next. Provide them with an email address contact that will be yours for the long term if your current student account will close after your graduation. Keep in touch periodically over the coming years—ideally more frequently than when you need another letter of recommendation for a fellowship or job application.

---

[6]Council of Graduate Schools, "April 15 Resolution: Resolution Regarding Graduate Scholars, Fellows, Trainees, and Assistants," http://cgsnet.org/april-15-resolution.

Many programs also accept students mid-year. Look at the deadlines and talk to faculty in those programs to determine when you should have your application submitted. From there you can adjust the timing discussed above.

In parallel to the graduate school application process you should also consider applying for graduate school fellowships. Also, unless you are independently wealthy or have a particular aversion to teaching, you should check all of the above if the application asks if you are interested in being considered for a teaching assistantship (TA), research assistantship (RA), and fellowship.

2.4.4   VISITING A GRADUATE PROGRAM YOU WOULD LIKE TO ATTEND

To make a well-informed decision, you should ideally visit the university and interact with the faculty and graduate students there. Many graduate programs organize visit weekends in the late winter/early spring. These are a great opportunity that you should try to take advantage of, if at all possible. You will have access to faculty and students on the visit and you will be able to see the facilities, campus, and community. Some programs invite only students that they have accepted into the program. Others will invite admissible students they would like to consider for funding offers.

If the programs you are interested in do not plan a visit weekend, you can arrange to visit on your own. The best point of initial contact would be the staff member in charge of the graduate program (e.g., program coordinator) or the faculty director of graduate studies (e.g., chair of the graduate studies committee). If you can't visit then you should make arrangements to set up virtual or phone conversations with the director of graduate studies and other faculty members you may be interested in working with.

You should think of a visit weekend like an interview. You are being interviewed, but you are also interviewing them. Everyone involved should be trying to determine if there is a good fit. Although you would not be expected to wear a suit, do present yourself professionally (business casual attire is usually appropriate). Be ready to present your experience and background clearly and succinctly. If you have engaged in undergraduate research, you may want to print out a few slides or have a copy of a research paper you wrote in order to share your prior experience more effectively.

Do your homework before you go on the visit. Learn as much about the university and faculty in the program as you can. If you are interested in working with a particular research mentor, become familiar with their recent research publications. Prepare questions that will help you determine if this is the right fit for you (see the list below).

"Questions to Ask Yourself and Others While Considering a Graduate Program"

This is a broad list of questions. Some of these questions are intended for you to answer yourself. Others you can find the answer to by exploring the university website. Some are questions you should ask of the faculty you speak with. Others you should ask of graduate students who are already in the program.

Overarching Questions to Ask Yourself

Am I most interested in experimental, computational, or theoretical research?

Would I rather be in an established group or do research with a more junior faculty member?

How much time and attention do I expect to get from my thesis advisor/research mentor?

Am I interested in interdisciplinary research and does this position fit with those interests?

Are the other students in the research group people that I can get along with?

School/City/Lifestyle

Is the campus a safe place? What safety programs are available (i.e., emergency phones, campus escorts)?

Is housing easy/difficult to find?

What are living expenses like?

Is there a reliable mass transit system?

Are there bike paths for commuting to campus?

What kinds of entertainment are available?

Will I be able to pursue the recreational activities I am interested in?

Do I feel comfortable in this community/area of the country?

Can I see myself living here for the next ∼5 years?

Program/General Atmosphere

What is the reputation of the program?

How is the quality of the teaching?

Are the required and elective courses ones that I am interested in taking? How frequently are they offered?

Are graduate students happy here?

How is the rapport among students, staff, and faculty?

How is the atmosphere for women and underrepresented minority students?

What is the office space policy for new graduate students?

Are the labs and facilities broadly accessible? How do I get trained to use these facilities?

Do faculty members collaborate on research or work separately? Is collaborative research encouraged and supported?

Funding/Financial Aid

Do I need to find/choose a thesis advisor before accepting an offer to join the program or do I have the opportunity to spend a semester or two on campus before I decide?

How do I apply for a teaching and/or research assistantship?

What fellowship opportunities are available from the program/university? Am I automatically considered for these opportunities or do I need to apply?

Is a tuition waiver included in my funding offer?

Is health insurance included in my funding offer?

What are the vacation/sick leave/family leave policies?

What is the stipend level? Do students live easily on this amount?

Does the funding continue through the summer months?

If I am offered an assistantship appointment, what are the work expectations?

What are the responsibilities associated with a teaching assistantship (TA)?

Is there training for new TAs?

How is my performance evaluated?

Who is my supervisor?

Who do I talk to if I need help with a problem in the classroom?

Research Mentor/Thesis Advisor

How stable is his/her research funding?

Does the advisor have tenure? If not, what is the tenure rate at this institution?

What is the advisor's reputation in the department?

How do the advisor's current students feel about working with this person?

Does the advisor treat students respectfully?

Does the advisor stand up for his/her students when a political situation arises?

Does the advisor give a lot of supervision or are students expected to work more independently?

How is one's thesis topic determined?

How is authorship handled on journal publications?

Will the research require traveling or working remotely?

How long does it usually take for the advisor's students to graduate?

Are there opportunities available to attend a conference or two each year?

Where have previous students gotten jobs?

2.4.5   GETTING ACCEPTED INTO A GRADUATE PROGRAM

Different programs will handle graduate applications differently. However, there is likely a committee that determines an applicant's overall fit for the program and selects the best applicants for broader circulation among the faculty members in the graduate program. For large programs and Master's programs that do not have funding associated with them, it is more likely to be a decision made at the committee level. For a Ph.D. program there is more match-making required because you will need to have an interest in the research taking place in a faculty members lab and they will need to have funding to support you as a research assistantship.

In many graduate programs there needs to be at least one faculty member who is interested in taking you on as an advisee in order for your application to progress. There are always exceptions though. Some programs have fellowship and teaching assistantship support that allows them to bring in more students without the promise of a research assistantship. And, students who have received a large external fellowship have more flexibility because they can often work with the faculty member of their choice without as much concern over the availability of funding for the research. I'll note, however, that the fellowships do not generally cover research expenses, so even a fully supported fellow is not "free" for the faculty research mentor. They will need to have the necessary funds to cover the expenses of the research and the time to provide research mentoring.

Getting Paid to Learn

Unlike some other disciplines, engineers are frequently given the opportunity to earn a stipend while doing research that will directly benefit their own degree progress. When I occasionally hear it taken for granted and expected that their education should be completely paid for, I shake my head in wonder at the entitled attitude of this individual. Having served in graduate school administration, I am able to state definitively that in many other fields of study graduate students must fund their education by working positions that have no bearing on their research progress and their work may not even be connected to their disciplinary expertise. In fact, having to pay for one's education in such away usually extends time to graduation dramatically.

As noted above, at some institutions your acceptance into the graduate program may be separate from an offer of assistantship funding. Be certain to understand the details of your particular situation before accepting an offer.

2.5   FUNDING OF RESEARCH

2.5.1   U.S. MODEL OF RESEARCH UNIVERSITIES

Students often come with misconceptions about where and how research funding is obtained. What students rarely appreciate is that research funding is very difficult to obtain. In most cases the funding for research (including a research assistantship) was obtained through a hard-fought and competitive proposal process. It is likely that their research mentor has spent an enormous amount of time and intellectual energy writing multiple proposals, of which only a subset is actually funded. The vast majority of research proposals that are written and submitted for consideration are rejected without being funded. Therefore, being supported on a research assistantship funded by a research grant is a privilege not an entitlement.

Student Perspective

"The thing I found most surprising about how research is conducted is the method by which most funding is procured and the overall attitude of researchers toward that source. When I first started learning about academic research, I expected budgets from research institutions to pay a large percentage of research costs. I believed that these budgets were heavily subsidized by student tuition and the earnings from previous research achievements at those institutions. This is not typically the case. Grants from the

> federal government are the single largest source of funding for the majority
> of universities and fields. Whether the funding is from a government agency
> such as NASA or the DOE, or from the Department of Defense, the money
> still comes from the American tax payer."

Grant funding may come from a federal source (such as the National Science Foundation or National Institutes of Health) or a private foundation (such as the American Heart Association or the Petroleum Research Fund). Research contracts are also a common funding source as well, and commonly come from federal sources (such as the Air Force Office of Scientific Research) or a private company (both small and large). Depending both on the source of the funding and the specific type of funding there may be very well-defined timelines and deliverables associated with the research. Some funding may require monthly, quarterly, annually, and/or final reporting associated with the project progress and outcomes. In other words, research funding comes with strings attached.

Given the overall framework of funding, I suggest to graduate students that they should treat their assistantship as professional employment. If you have an assistantship, you are being paid for your engineering skills through both the stipend (i.e., paycheck) and tuition (i.e., waiver of tuition). If you were working in industry, you would be expected to treat the job professionally, put in your best effort, and achieve regular progress. The same is expected in your graduate research.

2.5.2   FUNDING YOUR GRADUATE STUDIES

For graduate students in engineering, and particularly students pursuing a Ph.D. program, graduate school is usually paid for by a fellowship, a research assistantship, or a teaching assistantship.

> **Student Perspective**
> "I believed that you still had to spend lots of money to attend grad
> school. I am extremely pleased to know that through applying for fellowships
> and with how most engineering departments work, pretty much everything
> from living expenses to tuition and lab funding is potentially covered."

Fellowships come in many shapes and sizes. Some universities have fellowships to provide and others are available through external programs. A fellowship may provide a "full ride" that pays for all of your tuition and stipend expenses (for one or more years), or it may simply be a supplement to other types of assistantship funding. A full fellowship gives you a huge advantage because a potential research mentor does not need to find as much funding to support you. No graduate student is truly "free" because the research mentor must have the time to interact with you and be able to support other research expenses especially for experimental work, but it is *much* easier for a research mentor to take on a fellowship student than to find funding for an assistantship.

Fellowships provided by a university are usually ones that you are automatically considered for when you apply to the graduate program. The best way to ensure that you have a good chance at being considered for one of these is to have the best graduate school application possible and to submit it early. Do not wait until the deadline!! Many fellowship and assistantship opportunities will already be gone if your application is in the last batch of applicants to be considered.

There are also a variety of fellowships that you can apply for yourself as a senior undergraduate or a first-year graduate student. Your academic advisor or research mentor will be able to point you toward ones that might be a good fit for you, but you should consider looking into some of the following:

Department of Energy Computational Science Graduate Fellowship

Hertz Foundation's Graduate Fellowship

National Science Foundation Graduate Research Fellowship (NSF GRF)

National Defense Science and Engineering Graduate Fellowship Program (NDSEG)

National Defense Science, Mathematics and Research for Transformation (SMART) Scholarship

NIH Kirschstein-NRSA Individual Predoctoral Fellowship

Tau Beta Pi Association Graduate Fellowship in Engineering

As you progress through your graduate studies there are also additional fellowships available at later stages, particularly dissertation fellowships that are designed to help students finish up their Ph.D. program.

There are two basic types of support provided through universities that will fund your graduate studies. There were some variations on the specific titles depending on the institution, but many intuitions use the names research assistantship (RA) and teaching assistantship (TA). These assistantships usually provide for both tuition and a stipend for your living expenses. In return, you will be working on a research project or by teaching undergraduate students.

In many cases, research assistantships have a great deal of overlap with the research you will ultimately use for your thesis or dissertation. So, you are getting paid to do the research you would have needed to do anyway. Although the RA position may have a percentage appointment or certain number of hours associated with it, you will likely need to spend more time than what you are paid for in order to complete your degree in a timely manner. A good way to think about it is that you need to do a certain amount of research in order to earn your degree, and you are lucky enough to get paid for a portion of it!

As discussed above, there is more match-making needed in this case because you will need to be highly qualified, find a good fit between your research interests and a faculty member's research program, and have this match up with available funding support. Once you have identified schools that you are interested in attending, you also need to look at the research interests of the faculty members and contact them about the availability of funding. If they have an RA position available and you are a good match, then they may make you an offer!

In some cases graduate students may be brought into a degree program and initially funded by a teaching assistantship. In other cases, the TA opportunities may come later in the graduate experience and something that you do after you have progressed in your degree program. The type of work that a TA would do depends on the specific position and may include grading, holding office hours to answer student questions, running a discussion section, or teaching the lecture component of a course. Regardless of the position, there will be an instructor or faculty member in charge of the course, and you may also be working with other TAs on the same course.

Teaching assistantships, although excellent skill building opportunities, will not be as directly related to your degree progress. If you are interested in an academic career path, the opportunity to be a TA can help you gain invaluable experience. Even if you are not interested in being a faculty member some day, teaching a subject provides an opportunity for you to deepen your own understanding of it. If you are in front of a classroom for a portion of your TA work you will also be able to hone your presentation and explanation skills. Employers of every type appreciate these skills.

For students planning to pursue a Master's degree only, the funding opportunities are fewer. Sometimes RA and TA positions are available, but if you do not intend to continue on for a Ph.D. it is more likely that you will be paying tuition for the degree. Regardless, the investment in a Master's degree should pay off. On average, your salary will be higher,[7] your lifetime earnings with a M.S. vs. a B.S. are higher, and the unemployment rate is lower.[8] Employers are also increasingly requiring a Master's degree.[9]

Finally, there are student loans. Generally speaking, if you have student loans coming into graduate school, you will be able to defer your payment of them while you continue your studies. It's also often possible to get student loans for graduate studies to support the cost or supplement funding you have from the university.[10]

---

[7]Doubleday, J., 2013. Earnings Gap Narrows, but College Education Still Pays, Report Says, *Chronicle of Higher Education*, October 7.

[8]Council of Graduate Schools, 2013. "Open Doors with a Doctorate."

[9]Council of Graduate Schools, 2013. "Why Should I Get a Master's Degree."

[10]Council of Graduate Schools, 2013. "Financing Graduate Education."

2.5.3   FELLOWSHIP APPLICATIONS

As mentioned in the previous section, fellowships come in all sorts of shapes and sizes, from a "full ride" to a small supplement. However, there are a number of commonalities in the application process for those you would need to apply to yourself. This happens independent of the university you are applying to or attending, so you will need to manage those deadlines in addition to graduate school application deadlines. Look into these opportunities early. Although the deadlines vary quite a bit, many of them are due BEFORE the standard graduate school application deadlines.

As you look into each fellowship opportunity, carefully read the eligibility criteria. You will not want to waste time on an application where you do not meet the basic criteria or where you are not a good fit. Keep in mind that in some cases you will be applying as a senior undergraduate and in others as first year graduate student. Some fellowship competitions allow you to apply in more than one year as well.

Don't try to do this all on your own, without feedback. You will have a much higher likelihood of being successful if you plan ahead and seek out guidance. Determine if there is help available on your campus that will guide you in the fellowship application process. If there are workshops offered, seek these out and attend. There may be one-on-one help available if your campus has a writing center. You may also be able to seek feedback on portions of your application from an academic advisor or faculty member willing to read the essay portions. You should also use your network to find out if you know someone who has been successful in getting one of these fellowships. Being able to look at a successful fellowships packet will give you a model to emulate.

In addition to the fellowships available for your studies, there are also often small pockets of money that can help to defray other costs. Keep an eye out for other opportunities along the way, such as travel grants and other supplemental funding. Then later in your graduate studies, when you become a dissertator, look at fellowship opportunities again. Although there are not as many options as there are at the beginning of your graduate studies, in many fields there are dissertator fellowships that you can apply to which will help speed up your degree completion.

2.6   UNDERSTANDING THE ORGANIZATION OF YOUR RESEARCH GROUP

After you have joined a research group (or even while you are in the process of determining if a research group is a good fit for you), it is important to understand how the group is organized.

There will be research projects underway—some ramping up, some ongoing, and others winding down. You will be involved in at least one in detail, but you should also understand the basic themes of the other topics that your colleagues within the research group are engaging with. Having the basic framework of the research topics will allow you to sort and process additional information that you pick up in research group meetings, conversations with other research group members, and interactions with your research mentor.

> **Student Perspective**
> "My research group ... usually meets on a weekly basis to give updates on progress and get advice on how to proceed if we have a problem. I find this to be very beneficial because it helps me get a feel for what everyone else in my group is working on. Although it is hard to follow a lot of the time, it's good to learn what their projects are…"

Initially these interactions, particularly in research group meetings, may seem like a waste of time because nearly everything that is discussed is going over your head. But it is important to persist and try to follow as much of the information being exchanged as possible. You can also connect with one of the other more experienced students afterward to ask then to help you fill in some of the gaps. With time, you will be able, not only to understand more of what is being discussed, but also help provide useful feedback and ideas to the group yourself. Just keep in mind that it takes time to come up to speed, but you will make progress if you set goals for yourself that sometimes feel like a stretch.

> **Student Perspective**
> "Where I used to attend group meetings with glazed over eyes, I am now able [to] see what the other people are actually doing. However, I am usually not able to contribute too much because I still lack a significant amount of knowledge. Therefore, my main goal in the coming year is to be able to talk more in group meetings and provide the other group members with some helpful comments."

> **Student Perspective**
> "I think the most crucial element in my development during these meetings was that with every passing week, I felt more and more comfortable with the research, eventually to the point where I could try to suggest explanations and various solutions to problems in conjunction with the same inputs from the other members of the meeting. Having my ideas considered in a setting with three other people with considerably more experience in the field was very rewarding. The collaborative effort of people from different backgrounds to develop solutions to a problem or explanations for a phenomenon has become one of my favorite elements of research."

You may be paid to do research (for instance as a research assistant or as hourly pay) or you may be doing research for credit. Either way, it is likely that there is some type of funding supporting your salary and/or the purchases of resources that you need to conduct the research. You should understand what the funding source is for the research you are pursuing. It may be a federal grant, an industry contract, institutional funds that your research mentor has at their disposal, or some other mechanism. There may be multiple funding mechanisms supporting the various projects and people involved in the research group.

As a member of a research group, you also need to get to know the others engaged in the research group aside from your research mentor. Research groups come in many different sizes, from the small tight-knit groups to large international collaborations. There may be undergraduate researchers, graduate students, postdoctoral researchers, scientists, and faculty members. Your research group may also be collaborating with other research groups. These people may be working directly with you, using similar or complementary techniques, sharing research space with you, or they may be working at a different location or on a project that does not overlap with yours. Regardless, it is important to know who the research group members are and how they are connected to the work you are undertaking.

ASSIGNMENT 2-8:
INDIVIDUAL ASSIGNMENT – MAP THE ORGANIZATION OF YOUR RESEARCH GROUP

Create a visual depiction, or map, of the research you are currently working in (or planning to join). Talk with your research mentor and other lab members to understand what projects are underway, who are the people involved, and how the research is funded. You might depict one or more of the following.

- A diagram of the funded projects showing how they are interrelated, who is working on each, and what funding supports each person/project.

- For a highly collaborative group: this would include how the group collaborates with other individual researchers, research groups, and institutions across the ongoing projects.

- For an experimental group: the layout of physical lab space, how the experiments are organized, who utilizes on each piece of equipment, and how they projects/people are funded.

- For a computational group: the research projects that the group has going on and connections between the projects, people, and software being used/developed.

CHAPTER 3

Becoming a Researcher

3.1   DEVELOPING A RELATIONSHIP WITH YOUR RESEARCH MENTOR

Research groups can be set up in a variety of different ways and range in size from 1 to 100+. You may be working one-on-one with your research mentor or you may be working in more of a group setting where you meet with your research mentor along with others working on the same or related projects. In larger research groups you may find that there are researchers at a variety of different levels. This might include undergraduate students, graduate students, postdoctoral researchers, engineers, scientists, and faculty members. In some cases, your most immediate research mentor may be someone at a level just above your own. For instance, you may be an undergraduate researcher working most directly with a graduate student mentor.

Ultimately the responsibility for the research group, its direction, and the projects being pursued are determined by the lead faculty member or lead scientist/engineer, sometimes called the principal investigator or PI. This individual is also your research mentor (maybe you will think of this person as your Mentor with a capital M), but your interactions with this individual may be less frequent and may be in a group setting rather than one-on-one. You should not discount the others in the research group as they may provide you with invaluable information, advice, and mentoring that could prove to be important to your success.

Student Perspective

"I had some previous research experience at [a] National Lab. ... I had a mentor and a co-mentor that were constantly guiding me. I would meet with them several times per week to discuss how progress was going and ask [any] questions that I had. [My] two mentors also had offices right down the hall from mine and had an open-door policy so I could stop in and ask anything if I got stuck. This was so helpful to the ease and speed of my workflow. I could work on my project and when I ran into a problem, I would try to solve it on my own first, but if I couldn't figure it out, I could easily consult one of my mentors for help. Sometimes if they couldn't figure out the problem, they would point me in the direction of other researchers around the lab. This was a neat experience to draw on the expertise of researchers from different groups. I got to meet new people and learn about what they were working on while also getting a new perspective on the problem I was originally trying to solve. Prior to coming to grad school, I had guessed that my advisor would be play a similar role as my mentors at [the National Lab]. This semester has taught me otherwise. I didn't take into consideration the seemingly countless other obligations that grad school advisors have such as teaching, doing their own research, being active members of academic organizations which causes their time to be limited. Therefore, I do not have the same two-to-one relationship as I had at [the National Lab] which makes my work more independent. I think this is a good, and necessary step for me to take in my research career. This has made my problem solving skills much better and also has made me get to know the areas of expertise of the other students and staff members in my group. I'm learning who can possibly help me depending on the issue that I have run in to."

The Guides at Your Side

I would be hard pressed to count the number of mentoring relationships I have had over my career. Certainly somewhere in the multiples of hundreds, if I consider both those where I have been the mentor and those where I have been the mentee. These relationships have ranged from a few weeks to decades and have varying levels of involvement, but the common theme is a goal to help the other learn, evolve and be successful at what they are trying to accomplish. The more everyone understands the goals and motivations at the heart of a mentoring relationship, the more successful the results will be. This relies on communication and working to develop a rapport that will ultimately lead to a productive outcome.

Regardless of the size of the group and who specifically is your research mentor(s), you will need to take an active role in getting the mentoring you need to be successful with your research. Initially, you will be learning the basics of the project and the techniques you will be using, but even at this early stage you need to take ownership of your progress. Let your mentor know what you do know, and what you need help in learning, so that s/he can help you identify the resources that can assist you. As you gain more experience you are likely to be given more independence, both in terms of working more independently on specific tasks but also in carrying forward with the next steps before your next check-in with your mentor.

In the business world the term is called "managing up"—making the management of you as an employee easy for your boss—you can use these same ideas in a mentoring relationship by "mentoring up." In an article titled "Making the most of mentors: A guide for mentees," the authors[1] suggest that you take responsibility for the mentoring relationship by "guiding and facilitating the mentor's efforts." When working with a mentor, you have to figure out what you need from that person in terms of time, energy, and influence, and help that person to help you. Your goal is to ask for the help you need in a way that is easy for that person to give it to you. You may need other things from them—like letters of recommendation for a scholarship/fellowship for instance—and you need to make them aware of these needs as well as make it easy for them to meet your needs. Tell them about your goals, and where you want to go with you career. Tell them what would help you if you know and, if you don't, ask them what might help you to achieve your goals.

With your research mentor, determine how regularly will you meet—this may be more frequent at first and at critical points in the research or your degree process, so you may need to revisit and renegotiate the frequency of your interactions. If your mentor does not have regular meetings with you, take responsibility for requesting and scheduling these. Go beyond simply following through with the tasks that have been assigned to you and think ahead to what should come next, set goals that you can discuss, generate ideas for overcoming the research obstacles you have run into, and be responsive to the feedback you receive from your mentor. Most importantly, when you have an opportunity to interact with your research mentor, you should strive to be prepared.

- Have a clear plan, at least for the next step of your research.

- Be prepared to discuss what you have accomplished recently and what you plan to do next.

- Have questions to ask based on your research progress and/or your reading of the literature related to your project.

- Listen to your research mentor's responses, and write them down (either immediately or just after the interaction).

- Act on your plan and the suggestions made to you by your research mentor between now and your next interaction.

Student Perspective

"It was good to realize that the student is in some way expected and encouraged to dictate the schedule and flow of meetings. This made me more confident to meet with my professor and decide what an appropriate pace for my research is."

[1]Zerzan, J. T., Hess, R., Schur, E., Phillips, R. S., and Rigotti, N., 2009. Making the most of mentors: a guide for mentees. *Academic Medicine*, 84(1):140–144.

In my experience, the most effective and successful research students come to each meeting (whether it be in an individual or group format) with results in hand (either on a piece of paper or in a set of slides on their computer). They have thought about the results and what they mean, are ready to discuss them or ask questions about them, and have prepared a list of next steps that they will take. They take notes on what we discuss and what we decide to do (either in a lab notebook or a computer file). They identify resources they need, or questions that they have, so that I can help them move the research forward by pointing them in the right direction, connecting them to a person with the expertise they need, or purchasing something that is required for the research. These successful research students are also constantly keeping up with the literature, identifying recent publications that are relevant to their project. They bring those papers to my attention and they share relevant papers with other members of the research group. They also keep track of their own degree progress—deadlines for examinations, course requirements for the degree program, etc. In addition to sharing information with me, they share information with their peers, and mentor those who come in after them, either formally or informally. The reason these individuals are so successful is that they have taken ownership of their progress and help me to help them advance and succeed.

> **Student Perspective**
> "The change has been very gradual, but I'm starting to feel confident in my ability to understand the day-to-day research goals of the research group, and maybe more importantly, to know what questions to ask and when to ask them. This is a transition that I think many new researchers go through. How it often worked for me at first was that, when an unfamiliar topic came up, I doubted that I even had the technical background to have the means to learn about it. Not wanting to waste the time of the people who seemed to be familiar with the subject, I generally kept my questions to myself."

You can't expect to know everything when you setup into a new research project, but your goals should be to come up to speed quickly and ask relevant questions that will help you to obtain the background information you need. The worst thing to do is pretend you know something when you don't. Your research mentor, and the colleagues you work with, can't help you get to where you need to be if they don't know that you are lost. Phil Dee, who wrote the book *Building a Successful Career in Scientific Research*, highlights this as a foundational element providing "the ground rule" for your relationship with your research mentor: "communicate with your boss.[2]"

If you step back and think about it, you will see that academic research is a symbiotic relationship between the research mentor and the student. You and your research mentor must

---

[2]Dee, P., 2006. *Building a Successful Career in Scientific Research: A Guide for PhD Students and Postdocs.* Cambridge University Press.

depend on each other. In other words, it is mutually beneficial for you to be successful. My colleague, Prof. Irving Herman at Columbia University, wrote a somewhat tongue-in-cheek guide for graduate students in which he espoused the "The Laws of Herman.[3]" Several of the laws are about the symbiotic relationship mentioned between you and your research mentor, the last two being "Whatever is best for you is best for your advisor." and "Whatever is best for your advisor is best for you." Meaning that your success is to everyone's advantage.

Both Dee and Herman also bring up the topic of writing, which is a critical skill for every researcher at every level. If it is not something that you feel you are good at yet, don't worry, you will have many opportunities to practice and you will become better the more you write! If you take my advice above about preparing for meetings with your research mentor, you will automatically be writing something about your research. It may be in a bulletpoint list initially, but if you save these regular meeting notes you will find that later on, when you are at the stage of writing about your research, you can go back to these notes for reference and turn portions of your notes into sentences and even paragraphs. The other advantage of this chronological archive of information you have created along the way is that it can help you to refresh your memory about what you did to get to where you are, and the questions you were posing and answering. Although a thesis or journal article that you will write is *not* a historical recounting of every step and misstep that you took, a review of this information can help you to see the larger picture of your work.

ASSIGNMENT 3-1:
INDIVIDUAL ASSIGNMENT – REFLECTIVE WRITING ON GIVING AND RECEIVING FEEDBACK

Write a one-page reflection on giving and receiving feedback. First, describe a time that you received feedback that you (ultimately) found valuable. Discuss how you reacted to it at the time and how you looked back on this feedback as time passed. Then, describe a time that you provided feedback to someone else. Discuss the reaction/response you observed in the other person at the time and as time passed. Also discuss how you would have reacted if someone had provided you that same feedback in the same way.

[3] Herman, I. P., 2007. Following the Law. *Nature*, 445, 11.

ASSIGNMENT 3-2:
INDIVIDUAL ASSIGNMENT – REFLECTIVE WRITING ON DEE'S RULES

Consider the following "rules" from Phil Dee's chapter on "Choosing and Handling your Ph.D. Adviser"[4]:

> Rule 1: The ground rule: communicate with your boss.

> Rule 2: Keep your boss informed.

> Rule 3: Discover what makes your boss tick.

> Rule 4: Earn your boss's respect.

> Rule 5: Assert yourself.

> Rule 6: The golden rule: write for your boss.

Note that Phil Dee uses the term boss, where others use advisor, and this book most commonly uses research mentor. Although these terms can have different connotations, we are each talking about the same person (boss = advisor = research mentor).

Choose one of the rules above and discuss how you have seen this apply to your own research experience (or how you would expect it to emerge in a future research experience).

3.2   ALIGNING EXPECTATIONS

Some research opportunities and relationships come with clearly outlines expectations, but this is not always the case. When it is not discussed up front, it is up to you to seek clarification. Formalizing the relationships and the expectations can often be helpful. Many faculty are beginning to use written agreements with their students called variously a mentoring contract, mentor agreement, research agreement, or advising statement.[5,6] These can cover a wide range of topics, but the basic intent is for both he mentor and the mentee to understand the expectations of the relationship for the duration of the degree program.

---

[4]Dee, P., 2006. *Building a Successful Career in Scientific Research: A Guide for Phd Students and Postdocs*, Cambridge University Press.

[5]Branchaw, J., Pfund, C., and Rediske, R., 2010. *Entering Research: A Facilitator's Manual: Workshops for Students Beginning Research in Science*. WH Freeman.

[6]Masters, K. S. and Kreeger, P. K., 2017. Ten simple rules for developing a mentor—mentee expectations document. *PLoS Computational Biology*, 13(9). e1005709. https://doi.org/10.1371/journal.pcbi.1005709 and https://doi.org/10.1371/journal.pcbi.1005709.s001.

These expectations usually revolve around the topics of:

- shared goals including your career goals and what will be needed for you to achieve them;

- research skills you will need to develop to complete your project;

- work hours (number, time of day, days of week), work/life balance, and vacation time;

- graduate assistantship stipends, type of funding over time (e.g., RA vs. TA), and summer support;

- degree progress milestones and deadlines/goals for when they will be achieved;

- fellowship applications and grant writing assistance;

- expectations for documentation of research, publication, and authorship; and

- conflict resolution.

Traditionally the alignment of expectations has been either done more informally (or not at all). When it does occur informally, it likely happens over the course of time. Regardless of whether it is informal or formal, if your mentor does not embark on a conversation about these topics with you, it is something you will need to bring up. It can be anxiety provoking to be in the dark about what is expected of you. Having you understand your mentor's expectations will help you to meet them, but equally important, having your mentor understand your goals will help you to achieve them.

Student Perspective
"Over the meetings I've had with my research mentors, I've learned that the expectations they have for me and skills they suggest I work on developing seem to be centered around the idea of taking the time necessary to carry out my research carefully."

3.3 DEVELOPING EXPERTISE

In your pursuit of a research career, you will be transitioning from a novice learner to an expert in your chosen area of focus. But we should consider what is meant by the term expert. An expert is not someone who knows all the answers. An expert has significant knowledge on a topic, appreciate which knowledge is applicable in the given situation, and can seemly exert little effort in solving a problem. An adaptive expert is someone who approaches a new situation flexibly, applies their existing knowledge and skills, but is always seeking to learn more. It is important to recognize that experts must be lifelong learners to maintain and strengthen their expertise.

The U.S. National Research Council undertook an effort to link the research and practice on the topic of learning which culminated in a seminal book titled *How People Learn*[7] (cited well over 22,000 times in the literature). Key principles they summarize on the topic of *how experts differ from novices* include the following.

- "Experts notice features and meaningful patterns of information that are not noticed by novices."

- "Experts have acquired a great deal of content knowledge that is organized in ways that reflect a deep understanding of their subject matter."

- "Experts' knowledge cannot be reduced to sets of isolated facts of propositions but, instead, reflects contexts of applicability: that is, the knowledge is "conditionalized" on a set of circumstances."

- "Experts are able to flexibly retrieve important aspects of their knowledge with little attentional effort."

Everyone is a Novice

We are all novices and experts, it just depends on the topic area. I am still in the novice learner stage when it comes to baseball. I know the basics of the game, but when a fast and complex play occurs, I don't always follow exactly what happened and can't begin to predict the outcome. Happily, when I'm at one of my son's games, other parents watching the game with me are more expert. They are happy to explain what happened so that I can further develop my knowledge of the game. I won't ever have the expertise of a long-time player, but I'm working on becoming a more expert fan!

For instance, expert mathematicians notice patterns and identify classes of problems in order to develop an approach to a solution. They have not only solved many problems before, but they have also stepped back to consider the underlying principles of each problem and how solutions can be classified. This is the opposite of what I often see in novice engineers—they are often too quick to throw down an equation and immediately start plugging in numbers. As a student begins to refine and improve their approach, they find that they are most successful when they first look at a problem and think about what category of approach might work best, then work with the appropriate equations and manipulate them, and finally, at the end, plug in values and find a numerical solution. When using this more advanced approach, students

---

[7]National Research Council, 2000. *How People Learn: Brain, Mind, Experience, and School–Expanded Edition*, National Academies Press.

find that the parallels which can be drawn between problems become more obvious because the patterns become more recognizable.

I advocate that students start developing their intuition about problem solving in a new area of learning by developing an initial "guess" associated with specific problems—Do you expect it to be positive or negative? What magnitude would it be? What units? Then at the end of the problem, you check your solution back against the guess. If they agree, and the answer is confirmed, then you can build confidence in your intuition. If they don't agree, then either your guess was off or you made an error in your solution. If your guess was wrong, the solution process may shed light on where your intuition was off. If you feel confident in your guess, then it may help you to identify where you made an error in your solution. The process of thinking about the problem up front, and the retrospective analysis of the solution, will help you to advance toward more expert thinking. This does not just apply to coursework, it applies to research as well. You should have a hypothesis (a guess) before you begin, and you should design your research to explore that hypothesis to prove or disprove it.

Whether it is coursework or research, it takes an investment of time to develop your skills and begin to work toward expertise. Time on task has a big impact. But you must seek more than superficial knowledge. You need to develop expert knowledge that is both conditionalized on the context and centered around big ideas.[8] As you build expertise in an area you will notice that your performance will become automatic and fluid.

ASSIGNMENT 3-3:
INDIVIDUAL ASSIGNMENT – REFLECTIVE WRITING ON EXPERTISE

Consider areas of expertise that you have observed in others (such as your research mentor, others in your research group, classmates, etc.). Reflect on areas of expertise that you are developing with respect to your research. What can you do to either deepen your expertise in one of these areas, or develop expertise in an area you recognize as important to your research?

3.4   DEVELOPING YOUR OWN IDENTITY AS A RESEARCHER

As you engage in your academic career and through your experience as a researcher, one of your goals should be to become an independent, critical thinker. In your academic pursuits (and in life) you are on a continuous journey of learning. This journey is facilitated by self-awareness, reflection, and authentic experiences that will prepare you for where you will go next.

---

[8]National Research Council, 2000. *How People Learn: Brain, Mind, Experience, and School–Expanded Edition*, National Academies Press.

Self-awareness means knowing your own strengths and weakness, knowing what excites you and makes you curious, and knowing how you handle yourself when faced with both success and failure.

Reflection involves you taking the time to think about things that happen to you in life. How did you handle a challenge? How did you react to praise or criticism when it was given? How might you do things differently the next time? Reflection can be done all inside your own head, by journaling your experiences, or by talking to others in thoughtful conversation.

Authentic experiences are real-world activities that give you an actual taste of what it is like to do something. Sometimes this can be accomplished through a class assignment, but most often this means getting out into the professional world and trying your hand at something. In addition to research experiences, other valuable authentic experiences include internships for a company or national laboratory, and volunteer opportunities like Engineers Without Borders. What you will gain from these experiences will not only be technical experience, but also knowledge about yourself; who you are and who you want to be.

Becoming who you want to be can be thought of as *self authorship*. Marcia Baxter Magolda defines the term self authorship, or internal identity, as "simultaneously a cognitive (how one makes meaning of knowledge), interpersonal (how one views oneself in relationship to others), and intrapersonal (how one perceives one's sense of identity) matter.[9]" As you develop as a person and as a researcher, you will rely more on yourself for interpreting data rather than the interpretation of others; you will also begin to interact as a junior colleague, rather than a student with your peers and research mentor(s), and you will begin to develop your own identity as a researcher. Inherent in becoming a self-authored individual, and critical to your success as a researcher, will be your ability to realize that "the complexity of the world simultaneously requires systematic thinking, the ability to judge knowledge claims offered by authorities, constructing convictions, and openness to new possibilities.[10]" All of this may seem a tall order at the moment, but moving from authority dependence to self authorship, whether it be professors or parents, is important in both your professional and private lives.

> **Student Perspective**
> "I think this is what's most important for engineers in their capacity for self authorship. Through their education, they are able to form their own opinions, think critically, and problem solve. However, this means nothing

[9]Baxter Magolda, M., 1999. *Creating Contexts for Learning and Self-Authorship: Constructive–Developmental Pedagogy*, Vanderbilt University Press, p. 10.

[10]Baxter Magolda, M. and King, P. M., 2004. *Learning Partnerships: Theory and Models of Practice to Educate for Self Authorship*, Stylus, Sterling, VA, p. 3.

> if they are unable to share these opinions, listen to others, or form lasting
> personal and professional relationships."

The ultimate goal of higher education is to produce learners that have the following capacities.[11]

- *"Cognitive maturity*, characterized by intellectual power, reflective judgment, mature decision making, and problem solving in the context of multiplicity.

- An *integrated identity*, characterized by understanding one's own particular history, confidence, and capacity for autonomy and connection, and integrity.

- *Mature relationships*, characterized by respect for both one's own and others' particular identities and cultures and by productive collaboration to integrate multiple perspectives."

ASSIGNMENT 3-4:
INDIVIDUAL ASSIGNMENT – REFLECTIVE WRITING ON YOUR DEVELOPMENT AS A RESEARCHER

An important part of becoming a mature, independent researcher is discovering yourself. What interests and excites you? What motivates you?

Write a two-page self-evaluation on your development as a researcher. Reflect on where you have been, where you are now, and what you will work on next in your development as a researcher.

Revisit this assignment periodically. First complete the writing assignment described above, and then look back on previous assignments to remind yourself of where you have been and how you are developing.

3.5   TRACKING YOUR DEVELOPMENT AS A RESEARCHER

Initially, it may not seem relevant to track your progress in research, but in the long run it can prove exceptionally helpful. In particular, if you have some long term goals in mind (like submitting a research paper for publication, earning your Ph.D., or getting a faculty position) you can break down the larger goals into smaller steps that you will need to take, and track your progress along the way.

---

[11]Baxter Magolda, M. and King, P. M., 2004. *Learning Partnerships: Theory and Models of Practice to Educate for Self Authorship*, Stylus, Sterling, VA, p. 6.

Student Perspective

"With this being my first semester in a lab, it has been a large learning curve and looking at it now, it really puts into relief all the skills I need to further develop. I started out having to learn the safety procedures, location of everything in the lab, and other basics. Whether it was following the pipette rules and maintaining a clean working environment, it was all part of the learning curve. Evaluating now, it is evident the skills I need to develop. My basic laboratory skills are quite sufficient, but there is a large amount of equipment I will need to know how to use."

As a researcher, there are a variety of things that you will need to focus on, and master, in the years to come:

- a knowledge of the discipline in general and your specific subdiscipline specialty;

- a basic understanding of, and experience in, the steps and techniques of engineering research;

- ability to employ the scientific habits of mind that engineering research requires;

- awareness of ethical, social, political, and economic influences on, and impacts of, engineering research;

- skills in written and oral technical communication; and

- skills in collaboration and teamwork.

An Individual Development Plan (IDP) can help you to make progress on several fronts. For example, the American Association for the Advancement of Science (AAAS) has developed on online tool called *My IDP* available at `http://myidp.sciencecareers.org/`. For early-stage researchers the tool is helpful for identifying your skills, interests, and values and providing you with career paths that may be a good fit for you. Additionally, this tool helps researchers at various career stages in goal setting in areas like skill development, project completion, and career advancement. The SMART goal strategy emphasizes creating goals that are "specific, measurable, action-oriented, realistic, and time-bound," hence the acronym.

Using the "SMART" Principle[12]

S—Specific—Is it focused and unambiguous?

M—Measurable—Could someone determine whether or not you achieved it?

[12]Goal-setting strategies for scientific and career success, Fuhrmann, C. N., Hobin, J. A., Clifford, P. S., and Lindstaedt, B., 2013. *Science*, AAAS, Dec. 3. `http://www.sciencemag.org/careerscareersresearch/2013/12/goal-setting-strategies-scientific-and-career-success`. Accessed January 2018.

A—Action-oriented—Did you specify the action you will take?

R—Realistic—Considering difficulty and timeframe, is it attainable?

T—Time-bound—Did you specify a deadline?

Achieving your goals will take investment of time, but you will eventually be able to see gains. You will begin to understand more of the seminar talks you attend and the journal articles you read. You will gain the ability to operate independently and more efficiently. You will begin to contribute new ideas to the research conversations you engage in. You may end up developing specialty expertise that others in your research group rely on. As you invest more time and intellectual energy in your research, you will be begin to see payoffs in terms of progress and recognition for your research accomplishments.

ASSIGNMENT 3-5:
INDIVIDUAL ASSIGNMENT – QUALITIES OF A SUCCESSFUL RESEARCHER

List ten qualities that you will need to be a successful researcher. How far along in your development are you in achieving these qualities? How can you go about developing these qualities further?

ASSIGNMENT 3-6:
INDIVIDUAL ASSIGNMENT – REFLECTIVE WRITING ON MENTOR FEEDBACK

Individuals grow accustomed to receiving and accepting feedback in different ways and may react differently to the feedback depending on who it comes from, the context in which it is provided, and our mood at the time of receiving it. Some of us are more effective at either or both giving and receiving advice than others, but we can all become better. Before asking for feedback on your work or your development as a researcher, prepare yourself to receive it openly. Receiving feedback can be easier to handle when you ask for specific feedback on an area you already know you want to improve on and put the feedback to good use right away.

If you have a research mentor, request a meeting to receive feedback regarding the self-assessment above. Chose at least one area where you believe you have demonstrated strength, and at least one area where you believe you need additional development that you are ready to undertake.

Prior to the meeting, consider how you can be open to receiving the feedback you are going to receive. After the meeting, write a one-page reflection about what you heard and your reaction to it.

Be sure to express gratitude for the time your mentor takes with you to discuss this topic.

ASSIGNMENT 3-7:
INDIVIDUAL ASSIGNMENT – SELF ASSESSMENT

Use the "Evaluation of Research Progress and Researcher Development" rubric in Table 3.1 to conduct a self-assessment. It covers a range for skills, from **Research Documentation** to **Stress Management**. Your specific research project will also require specific skills, so space is provided at the end of the self-assessment for you to define these and track your progress on their mastery.

Note that the skills are cumulative (from the left column, to the middle, and finally the right column). If you have **Mastery** in an area, you will have demonstrated the items listed under **Beginning** and **Developing**, as well as **Mastery**.

As you enter into research, it is likely that you are at the **Beginning** stage in most all areas. Take a look ahead to the next level and see what items you should be working on in your development. If you don't know how to make progress on this next level, it is likely that your research mentor will be able to give you some strategies for making progress.

Periodically assess yourself for your research progress and development as a researcher (pages 49–55). Consider sharing the assessment with your research mentor to prompt discussion about where you are in your development as a researcher and how you can make progress in areas you would like to improve in. Although you may be able to achieve mastery in some areas during your degree progress, other topics may be something you work on throughout your career.

3.6   BEING AN EFFECTIVE TEAM MEMBER AND COLLABORATOR

Being an effective team member requires a wide range of social and organizational skills. You may already come equipped with many of these skills given your prior experience, but there are likely areas in which you may need to gain experience or improve on your current capabilities.

One of the earliest aspects of being a good team member that you may encounter is the etiquette and expectations of participating in a team meeting. In the context of research, this comes up in research group meetings or lab meetings. There are several strategies that you can employ to determine the appropriate type and level of engagement expected of you. Another basic strategy is simply to ask the question of those who are already in the know. This can be posed to your research mentor, and you can also ask other research group members to give their impressions of the expectations. This may mean coming prepared with certain materials in

Table 3.1: "Evaluation of Research Progress and Researcher Development" rubric (*Continues.*)

Evaluation of Research Progress and Researcher Development		
Milestones and Timeline		
The ability to set realistic goals and use time and resources effectively; to obtain the maximum benefit from a minimum investment of time and resources.		
❏ **Beginning**	❏ **Developing**	❏ **Mastery**
Demonstrated by:	Demonstrated by:	Demonstrated by:
• focusing on tasks at hand without dwelling on past mistakes • completing assignments on time • making use of reference books and literature • coordinating and working with others on group project assignments • preparing for scheduled appointment times • using unscheduled time efficiently	• planning ahead • setting up an effective schedule • coordinating schedule with others • demonstrating flexibility • moving forward when mistakes are made • accepting responsibility in group activities • identifying alternative resources • using library and internet resources effectively • updating solutions based on review of available literature	• setting priorities and reorganizing as necessary • performing multiple tasks simultaneously • delegating when appropriate • following up on projects in a timely manner • managing meeting time effectively • considering professional goals in the context of project • demonstrating the ability to say "no" if requests made in conflict with set goals • actively seeking resources to solve problems or answer questions • using limited resources creatively
Responsibility		
The ability to fulfill commitments and to be accountable for actions and outcomes.		
❏ **Beginning**	❏ **Developing**	❏ **Mastery**
Demonstrated by:	Demonstrated by:	Demonstrated by:
• being punctual • completing tasks on time • following through on commitments • accepting responsibility for own actions and outcomes • recognizing own limits	• providing constructive feedback to the appropriate person(s) • offering and accepting help • completing projects without prompting • contributing to the provision of a safe and secure environment	• promoting education • accepting leadership roles • delegating as necessary

Table 3.1: (*Continued.*) "Evaluation of Research Progress and Researcher Development" rubric (*Continues.*)

Professionalism		
The ability to exhibit appropriate professional conduct and to represent the profession effectively.		
❏ **Beginning**	❏ **Developing**	❏ **Mastery**
Demonstrated by:	Demonstrated by:	Demonstrated by:
• following University, Department, and research group policies • demonstrating honesty, integrity, and respect to others • seeking opportunities for leadership • demonstrating an awareness of the professional role of the engineer in society	• participating in professional activities/organizations • identifying positive professional role models • discussing societal expectations of the engineering profession • awareness of the impact of ethical issues and legal issues on the engineering profession • acting on moral commitment	• acting in a leadership role • actively participating in professional organizations • actively promoting the engineering profession • advancing the engineering profession outside of the academic program

Commitment to Learning		
The ability to self-assess, self-correct, and self-direct; to identify needs and sources of learning; and to continually seek new knowledge and understanding.		
❏ **Beginning**	❏ **Developing**	❏ **Mastery**
Demonstrated by:	Demonstrated by:	Demonstrated by:
• identifying problems • identifying needs for further information • formulating appropriate questions • identifying and locating appropriate resources • attending class consistently • showing evidence of preparation prior to class • showing attentiveness • demonstrating a positive attitude toward learning • participating in small groups • offering own thoughts and ideas	• identifying own learning needs based on previous experiences • setting personal and professional goals • seeking new learning opportunities • seeking out professional literature • prioritizing information needs • reconciling differences in opinions or information • analyzing and subdividing large questions into components • demonstrating confidence in presenting material	• researching and studying areas where knowledge base is lacking • reading articles critically and understanding limitations • accepting that there may be more than one answer to a problem • recognizing the need to verify and then verifying solutions to problems • formulating and re-evaluating position based on available evidence • demonstrating confidence in sharing new knowledge

Table 3.1: (*Continued.*) "Evaluation of Research Progress and Researcher Development" rubric (*Continues.*)

Communication Skills		
The ability to communicate effectively (i.e., speaking, body language, reading, writing, listening) for varied audiences and purposes.		
❑ **Beginning**	❑ **Developing**	❑ **Mastery**
Demonstrated by:	Demonstrated by:	Demonstrated by:
• understanding and applying English (verbal, written, grammar, spelling, expression) • communicating appropriately per situation • providing appropriate feedback to team members and faculty • recognizing differences in communication styles • recognizing impact of non-verbal communication: maintaining eye contact, listening actively	• modifying communication when necessary • reflecting, clarifying, and restating messages • utilizing non-verbal communication to augment verbal messages • exhibiting appropriate communication per situation • maintaining quality in written work • maintaining quality in oral work • utilizing technology	• modifying written and verbal communication to meet needs of various audiences • presenting verbal or written messages with logical organization and sequencing • maintaining open and constructive communication • communicating professional needs and concerns • utilizing communication technology effectively

Interpersonal Skills		
The ability to interact effectively with faculty research mentor, scientific staff, graduate students, team members, and other department personnel, and to deal effectively with cultural and ethnic diversity issues.		
❑ **Beginning**	❑ **Developing**	❑ **Mastery**
Demonstrated by:	Demonstrated by:	Demonstrated by:
• maintaining attentive behavior • demonstrating acceptance of limited knowledge and experience • communicating with others in a respectful, confident manner • appropriate behavior in discussion • maintaining professional demeanor in interactions • respecting differences in others • recognizing impact of non-verbal communication	• seeking to gain knowledge and input from others • assuming responsibility for own actions • establishing trust and motivating others • recognizing impact of non-verbal communication and modifying accordingly • discussing problems with the appropriate person(s)	• approaching others to discuss differences in opinions • talking about difficult issues with sensitivity and objectivity • responding effectively to unexpected situations • delegating to others as necessary

Table 3.1: (*Continued.*) "Evaluation of Research Progress and Researcher Development" rubric (*Continues.*)

Use of Constructive Feedback

The ability to identify sources of feedback, to seek out feedback, and to effectively use and provide feedback for improving personal interaction.

❏ Beginning	❏ Developing	❏ Mastery
Demonstrated by:	Demonstrated by:	Demonstrated by:
• using active listening skills • showing a positive attitude • critiquing own performance • maintaining two-way communication • actively seeking constructive feedback and assistance	• assessing own performance accurately • seeking, accepting, and integrating feedback from others • developing a plan of action in response to feedback	• considering multiple approaches when responding to feedback • modifying feedback given to others according to their learning styles • engaging in non-judgmental, constructive, problem-solving discussions • reconciling differences with sensitivity

Critical Thinking

The ability to question logically; to identify, generate, and evaluate elements of logical argument; to recognize and differentiate facts, illusions, assumptions, and hidden assumptions; and to distinguish the relevant from the irrelevant.

❏ Beginning	❏ Developing	❏ Mastery
Demonstrated by:	Demonstrated by:	Demonstrated by:
• considering all available information • recognizing gaps in knowledge base • articulating ideas/problems • raising relevant questions	• understanding scientific method • critiquing hypotheses and ideas • formulating alternative hypotheses and ideas • examining new ideas • being able to distinguish relevant from irrelevant information • recognizing fact vs. opinion	• exhibiting an openness to contradictory ideas • assessing issues raised by contradictory ideas • justifying selected solutions • determining effectiveness of applied solutions • identifying complex patterns of associations • demonstrating intuitive thinking • distinguishing when to think intuitively vs. analytically • recognizing own biases and suspending judgmental thinking • challenging others to think critically

Table 3.1: (*Continued.*) "Evaluation of Research Progress and Researcher Development" rubric (*Continues.*)

Scientific Literacy		
The ability to use processes and skills of science to conduct investigations; to recognize and define problems, analyze data, develop and implement solutions, and evaluate outcomes.		
❒ **Beginning**	❒ **Developing**	❒ **Mastery**
Demonstrated by:	Demonstrated by:	Demonstrated by:
• recognizing problems • identifying questions • knowing the basic steps of the problem-solving process (stating the problem, describing known solutions, identifying resources needed to develop solutions, beginning to examine multiple solutions to the problem) • seeking to fill gaps in knowledge • understanding differences between primary, secondary and other sources	• distinguishing between fact and hypotheses • applying the problem-solving process • prioritizing problems • consulting with others to clarify the problem • identifying contributors to the problem • accepting responsibility for implementing solutions • considering consequences of possible solutions • generating alternative plans when difficulties or obstacles present themselves	• forming possible solutions • designing a data collection scheme and collecting data • drawing conclusions about the validity of the possible solution • seeking alternative hypotheses and contradictory ideas • evaluating outcomes • reassessing solutions
Research Documentation		
The ability to effectively document research approach, progress, hypotheses, and outcomes.		
❒ **Beginning**	❒ **Developing**	❒ **Mastery**
Demonstrated by:	Demonstrated by:	Demonstrated by:
• recording research findings • identifying methods used	• keeping record of research progress • writing out steps to possible solution • providing documentation that others can follow	• describing thought processes, hypotheses and outcomes • supporting methods chosen with literature references • using project managements tools to stay on task

Table 3.1: (*Continued.*) "Evaluation of Research Progress and Researcher Development" rubric (*Continues.*)

Stress Management		
The ability to identify sources of stress and to develop effective coping behaviors.		
❏ **Beginning**	❏ **Developing**	❏ **Mastery**
Demonstrated by:	Demonstrated by:	Demonstrated by:
• recognizing own stressors or problems • recognizing stress or problems in others • seeking assistance as necessary • demonstrating appropriate responses • maintaining professional demeanor	• accepting constructive criticism appropriately • handling unexpected changes appropriately • maintaining balance between professional and personal life • establishing outlets to cope with stressors	• recognizing when problems are unsolvable • demonstrating a preventive approach to stress management • offering solutions for stress reduction • assisting others with stress • establishing a support network • prioritizing multiple commitments • tolerating inconsistencies • responding calmly to urgent situations

Project-Specific Research Skill 1		
❏ **Beginning**	❏ **Developing**	❏ **Mastery**
Demonstrated by:	Demonstrated by:	Demonstrated by:

Table 3.1: (*Continued.*) "Evaluation of Research Progress and Researcher Development" rubric

Project-Specific Research Skill 2		
❏ **Beginning**	❏ **Developing**	❏ **Mastery**
Demonstrated by:	Demonstrated by:	Demonstrated by:

Project-Specific Research Skill 3		
❏ **Beginning**	❏ **Developing**	❏ **Mastery**
Demonstrated by:	Demonstrated by:	Demonstrated by:

advance, listening attentively and asking for clarification at key points, asking critical questions about the topic at hand, or contributing to the discussion by giving thoughts and ideas. It is better to ask in advance of the meeting what the expectations are so you are not unprepared, however you can simply choose to attend a meeting or two and use your observation skills to take careful note of how the interactions work and who is expected to speak and about what topics. It is highly likely that the expectations for your participation initially will be much lower and will increase with your experience and time within the group. As you get more comfortable in the group you may find that you are talking more. However, always be certain that your contributions are succinct, meaningful, and on topic so that the meeting is maintaining forward momentum and you are not being wasteful of other people's time. Although you may feel "out ranked" because of your limited experience, don't discount the insights that can come from a new person's prior experience or simply a fresh set of eyes on the topic.

In general, you want to be respectful of your research mentor's and other team members' time. This means that you should arrive on time when a meeting or event has been scheduled, and you should come prepared. In the case of a one-on-one meeting with your research mentor, this means having thought through what outcomes you would like to get from the meeting, as well as what your research mentor will expect to learn from you about your progress. If you will be presenting a literature review or some results from your own research, you will want to have this material organized and ready to discuss. If you will be needing to use audio visual equipment, you will want to arrive at the room in advance and have everything set up and ready to go so that meeting time is not wasted while you're trying to get the equipment to work.

Outside of meetings you are likely to have a variety of different kinds of interactions with other team members. If you need help from someone, it is perfectly reasonable to ask for it, but you should strive to make it as convenient as possible for them to provide it to you. If you will be trained on a piece of equipment or a technique, ask if there is information that would be helpful for you to read in advance so that you can come better prepared. When you are being trained, give it your undivided attention and take notes. Ask clarifying questions and for repetition if necessary, so that you can minimize the likelihood that you will need to return again for repeated training.

> **Student Perspective**
> "There was a very frustrating period of time where I wasted a lot of time trying to make the experiment work with my limited knowledge … when all I had to do was ask a grad student for advice. Back then, I often forgot that scientific research is a collaborative effort and that asking others for help can often save you a lot of time."

Often collaborative research requires interdependencies between your research outputs and the input needs of others in your research group. It can be complicated to move the research forward in an efficient manner. A high level of communication is required so that each researcher understands exactly what is needed and what is being promised. It is also important to have a clear understanding of the time frames in which activities will be taking place. What you might consider to be a small delay could negatively impact someone else's later progress more significantly. A shared calendar, project management timeline, and/or scheduling software can assist in making certain that everybody is aware of the timing of a particular project milestones and deliverables so that the research can be kept on track and moving forward. Although your research mentor may provide this for you, if it is not already available you should consider consulting with your research mentor, and other group members, about putting something in place that would be helpful to everyone.

As your engagement in research progresses, you may need to lead a meeting. That may be an informal meeting between members of your own group to resolve an issue, or a meeting with members of another research group to coordinate efforts in a joint project. Whether formal or informal you should come to the meeting with clear goals in mind and the order in which you would like to address topics. A written agenda is often helpful. Taking time to prepare an agenda in advance can make the meeting run smoothly and ensure that you will accomplish your goals. It also allows everyone to see the plans for the meeting and make any needed adjustments to the order and topics up front.

Facilitating a productive and respectful discussion can sometimes be challenging. Although it may seem too formal to start the meeting by agreeing to a set of ground rules, you can insert them into the meeting when needed. For instance, if the conversation is going too far off topic, you can bring people back by saying something like "that's interesting, but we have limited time today so we'll need to stick with our agenda to get done." When you are leading a meeting one of your responsibilities is to ensure that everyone has an opportunity to provide input and express their opinions. If one of the group members is getting talked over, or ignored, you can say "we need to be sure we hear from everybody on this topic, let's go around the room and get each person's input." There are numerous ways to be effective, one strategy is to watch others who are effective where you want to build skill and emulate them.

Being an effective team member also means getting to know the others on the team. A basic understanding of the other individuals you interact with can help to reduce friction and avoid conflict. For instance, something simple like asking about people's music preferences before playing your favorites at high volume in the lab can help to avoid irritation of other group members. Or if you are always requesting to meet in the early evening when a team member needs to pick up a child at day care, you could come across as insensitive when you had no intention of doing so.

Even if you do take preventive measures, conflict can still come up. Rather than avoiding or ignoring the situation, you can often achieve a better result by addressing the issue sooner rather than later. Approach the individual or group with openness and seek to understand the issue. It is likely that your effort will be appreciated, and you can work together to find an appropriate resolution.

3.7   WORKING WITH A DIVERSE RESEARCH TEAM

Depending on your prior experience and how well aligned it is with the atmosphere of the research group that you are joining, you may find that you have some adjustments to make as you begin to engage in a new research project. You may come to research with the idea that you will be the "lone genius" who operates entirely independently. This is exceptionally rare, and not particularly realistic when one is just beginning to engage in research. Most engineering research is conducted in a team environment. It may be a team of two—you and your research mentor—but more often it is a team of several or many. The team usually includes a faculty member (or members) and graduate students. Many also incorporate undergraduate researchers, postdoctoral researchers, and scientists. These people may all be in the same building at the same institution, they may be spread across a campus, or they may be distributed at different institutions across the country or even throughout the world. There are good reasons for this. Teams of people are able to tackle more complex and broader reaching research problems.

As a result of how research is organized, this inevitably means that you will need to work with others effectively. Not everyone in your research team will come to the group with the same background and experiences. If you think about even the most apparently homogeneous group of people you have interacted with, you can identify ways in which the group is diverse—for example the people in the group may look like each other but they may practice different religions, identify with different political groups, or spent their childhoods being raised in different environments. Each one of these differences gives the group broader experiences to draw from, and if it is a group of engineers this diversity may influence the way in which design decisions are made or research problems are posed. Ideally, we would strengthen the diversity of our engineering work groups to include people from a wide range of different backgrounds, and have diversity along many other spectrums, such as gender, race, etc. Companies recognize and hire with diversity in mind because research has shown that diverse groups are more productive, creative and innovative.[13] This is true for engineering research environments as well and engineering design. We all benefit from the higher-quality ideas—in terms of feasibility and effectiveness—that are produced by diverse groups and the critical analysis of alternatives when a wider variety of viewpoints is discussed.

---

[13]Women in Science and Engineering Leadership Institute, "Benefits and Challenges of Diversity," University of Wisconsin–Madison, http://wiseli.engr.wisc.edu/docs/Benefits_Challenges.pdf.

Student Perspective

"Education has also taught me a great deal about relationships with other people. Specifically how to work with others that may not share the same viewpoint as your own. Particularly in the field of research, tolerance of everyone's ideas is critical for success."

In order to build and maintain an effective diverse team we need to recognize some things about human nature. Whether we like it or not, we all carry unintentional biases (also called implicit biases) that are "habits of mind" and are influenced by where we have grown up and spent our lives. Harvard University psychology researcher Prof. Mahzarin Banaji was quoted as saying "Implicit biases come from the culture. I think of them as the thumbprint of the culture on our minds."[14]

As an example, if someone is asked to list the stereotypical characteristics of a man, they'll come up with many of the following: tall, physically strong, respected, intelligent, has high status, leaders, sexist, like sports, providers, aggressive.[15] However, even though we can list these stereotypes (women and men carry the same stereotypes in their mind about women and men) it does not mean we believe all men have these characteristics. We know that any individual man does not embody all, or even most, of these and I am certain that we could find some men who don't display any of the characteristics on the list. Similarly, the stereotypical characteristics of women can be listed: emotional, caring, soft, care about appearance, talkative, small built/petite, submissive, dependent, motherly, feminine.[16] But again, we don't expect that every woman we meet will conform to these characteristics. And it is not just gender at play. We hold numerous biases about all sorts of things like race, ethnicity, age, country of origin, etc.

The problem comes when we make quick decisions or have limited information. When we do this we fall back on our stereotypes. Say there is an election for county sheriff and all you know about the slate of candidates is that one has a male name and the other has a female name. The responsible thing to do would be to not vote without knowing more information, but many people will vote and with such limited information the stereotypes may have influence: we tend to think of police officers needing to be physically strong and in the role of sheriff they would have to serve as a leader. These are two characteristics we more readily associate with men than with women. These associations could push the voter toward the male candidate, even though we know nothing about the actual qualifications of the two individuals running in the election.

[14]Hill, C., Corbett, C., and St. Rose, A., 2010. *Why so few? Women in science, technology, engineering, and mathematics*, American Association of University Women. Washington, DC.

[15]Ghavami, N. and Peplau, L. A., 2013. An intersectional analysis of gender and ethnic stereotypes: Testing three hypotheses. *Psychology of Women Quarterly*, 37.1, 113–127.

[16]Ghavami, N. and Peplau, L. A., 2013. An intersectional analysis of gender and ethnic stereotypes: Testing three hypotheses. *Psychology of Women Quarterly*, 37.1, 113–127.

Unfortunately, these issues of unconscious bias play out in subtle ways that can have big impacts: who gets hired for a job,[17] who gets the award,[18] who gets the grant funding.[19] Because most of us would want the most qualified person to get the job, the student with most potential to get the fellowship, and the best idea to get the grant funding, we need to be aware of our biases and work against applying them unintentionally.

> **Student Perspective**
> "[Thinking] about our inner biases and how they influence our lives and decisions was awakening. How easily we can form biases based on misinformation and then base judgments on those facts and then follow it by the act of actually defending our biases was a good realization."

The first thing to recognize is that you are not a bad person because you have biases. Everyone has them. What we all need to do is to recognize our own biases and work to overcome them. Some useful strategies are as follows.[20]

Recognize and Replace: Become more aware of the biases that you carry and work to replace them by thinking of counter examples. The research shows that it is fruitless in the long run to simply try to repress stereotypes—this backfires.[21] Challenge your automatic thoughts with concrete examples. Visualize an engineer. Now visualize someone you know, who is an excellent engineer and also belongs to a group that is underrepresented in engineering.

Intergroup Contact: Much of our work as engineers is done collaboratively and in teams. Get to know the other research group members as individuals.[22] Challenge your assumptions of who they might be given the stereotypical information available on the surface. Pay attention and don't dismiss information that does not fit with the

---

[17]For example, see Segrest Purkiss, S. L., Perrewe, P. L., Gillespie, T. L., Mayes, B. T., and Ferris, G. R., 2006. Implicit sources of bias in employment interview judgments and decisions. *Organizational Behavior and Human Decision Processes*, 101.2, 152–167.

[18]For example, see Lincoln, A. E., Pincus, S., and Schick, V., 2009. Evaluating science or evaluating gender. *American Physical Society News*, 18.8.

[19]For example, see Ley, T. J. and Hamilton, B. H., 2008. The gender gap in NIH grant applications. *Science*, 322.5907, 1472–1474.

[20]Adapted from Carnes, M., Fine, E., Romero, M., and Sheridan, J. Breaking the bias habit, Women in Science and Engineering Leadership Institute (WISELI), University of Wisconsin–Madison, `https://wiseli.wiscweb.wisc.edu/workshops/bbh-inclusive-campus/`; see also Carnes, M., Devine, P. G., Manwell, L. B., Byars-Winston, A., Fine, E., Ford, C. E., Forscher, P., Isaac, C., Kaatz, A., Magua, W., Palta, M., and Sheridan, J., 2015. The effect of an intervention to break the gender bias habit for faculty at one institution: A cluster randomized, controlled trial. *Academic Medicine*, 90(2):221–30.

[21]Macrae, C. N., Bodenhausen, G. V., Milne, A. B., and Jetten, J., 1994. Out of mind but back in sight: Stereotypes on the rebound. *Journal of Personality and Social Psychology*, 67(5), 808.

[22]Pettigrew, T. F. and Tropp, L. R., 2006. A meta-analytic test of intergroup contact theory. *Journal of Personality and Social Psychology*, 90(5):751–83. And Lemmer, G. and Wagner, U., 2015. Can we really reduce ethnic prejudice outside the lab? A meta-analysis of direct and indirect contact interventions. *European Journal of Social Psychology*, 45(2):152–68.

stereotype. Appreciate the strengths that they bring to the shared goals your research group is working toward.

Model Inclusion: Use inclusive language. When a joke is inappropriate, don't laugh. Approach students who may be different from you and get to know them. Don't always interact with the same people; mix with others and get to know them better.

Perspective Taking: Develop your ability to take someone else's perspective and see the world through their eyes.[23] Use your empathy skills to see their perspective.

ASSIGNMENT 3-8:
INDIVIDUAL ASSIGNMENT – REFLECTIVE WRITING ON PERSPECTIVE-TAKING

Practice perspective-taking. Think about your early experiences with engineering research:

What did you feel like walking into the first research group meeting or research seminar? What were your first impressions about the people? What do you think people assumed about you?

Now consider someone who would be in this same situation who is from a background or group other than your own (e.g., different gender, race, ethnicity). How do you think it would it feel for that person to walk into their first research group meeting or research seminar? Write a 300–500 word reflection on this topic.

ASSIGNMENT 3-9:
INDIVIDUAL ASSIGNMENT – CASE STUDY

Instructions:

Read the brief case description provided. Reread while noting the important information and questions that are raised in your mind about the information provided, the individuals involved, and their situation. Determine both the basic issues and any deeper underlying issues at play. Consider the questions posed at the end of the case and how you would respond to these questions, as well as other questions that could be asked of this case. Write a one-page response that includes a brief summary of the case and its issues, your answer to the questions posed, and recommendations based on your understanding of the situation posed in the case.

---

[23]Todd, A. R., Bodenhausen, G. V., Richeson, J. A., and Galinsky, A. D., 2011. Perspective taking combats automatic expressions of racial bias. *Journal of Personality and Social Psychology*, 100(6), 1027.

Case description:

Jeff is excited to begin his research career and join Prof. Jones' research group. Now that he has moved to town and gotten settled he's eager to start getting acquainted with the other research group members. He has just been assigned a desk and has begun to meet the other students in the group with the help of Sam who Prof. Jones has asked to show him around. While he and Sam are talking with other students in the group, Jeff notices another student come into the room and go to a desk in the corner without greeting anyone. Later as Sam is showing him around the building Jeff realizes that he did not get introduced to the student at the corner desk and asks Sam who she is. Sam replies "Oh, that's Ellen," and continues talking about other points of interest in the building.

After working in the groups for a couple of weeks and seeing Ellen come in every day and go to her desk in the corner without any interaction among the group, Jeff decides that there must be some issue with Ellen and proceeds to ignore her like the rest of the group members.

Questions to consider:

Should Jeff introduce himself to Ellen even though neither Sam nor Ellen have initiated an introduction?

Is it appropriate for Jeff to inquire with other research group members about Ellen to find out more about the situation?

How can Jeff go about engaging with Ellen given that they are both members of the same research group?

Would your answers above change if the person's name was Shaheed instead of Ellen? How do stereotypes play into your answers for Ellen, and for Shaheed?

3.8   DEVELOPING GLOBAL COMPETENCY

One important way to broaden your horizons and become a more culturally competent engineer is to spend time in another country. Ideally, you will do this in an immersed way and with a structured program. Study abroad is an option, both as an undergraduate and as a graduate student. There are also opportunities to work abroad and conduct research abroad. And of course, you can attend conferences in other countries. The more time you are able spend, and the quality of your immersion in the culture of that county, the more it will increase the impact of the experience.

> **Student Perspective**
>
> "There is a fairly evident way in which my interpersonal foundation is lacking as a result of my education and more specifically its location. I grew up in a community that was predominantly white with some amount of people of Asian descent. This has to some extent limited the variety of cultures to which

I have been exposed. This means that there are many cultures in which I still have something of a gap to cross to develop mature relationships. The global nature of research will allow me to have contact with many more cultures and start to understand them so that I can continue to work on my interpersonal foundation of self-authorship."

Increasing your cultural competence will benefit you in the long run regardless of your career goals. The world is more interconnected than it has ever been and the field of engineering is inherently a global one. Research teams are becoming more global and international collaborations commonplace. As a result, employers are interested in hiring individuals with the skills to operate in a range of settings and with people form a variety of backgrounds.

What does an immerse experience give you—both positive and negative? There are challenges in navigating a new culture and place. You will have experiences that stretch you a bit and force you to be more flexible and adaptable. You will also need to be self-reliant, and it will enhance your independence. You will learn about yourself through your experiences with the culture you are immersed in. Marcia Baxter Magolda connects it to the *self authorship* (identity development) ideas discussed previously: "Intercultural maturity includes the ability to use multiple cultural frames to construct knowledge, engaging in meaningful relationships with diverse others that are grounded in appreciation of difference, and the capacity to openly engage challenges to one's beliefs.[24]" Particularly as you work with people from other cultures, it is critical that you are able to see ideas and events from more than just your own perspective. "Mature relationships are characterized by respect for both one's own and others' particular identities and cultures as well as by productive collaboration to negotiate and integrate multiple personal needs.[25]"

Student Perspective

"Interactions and team-work with classmates and teachers coming from various linguistic, cultural and religious backgrounds has made me understand the intricacies of society, a way to relate to people and build relationships both personal and professional."

Some humorous illustrations of how cultural differences can impact both understanding and ability to work together came out as a series of commercials (adverts) from HSBC, a large international banking and finance origination. They illustrate a few cultural differences around the world with a bit of humor thrown in. In one of their ads they show a British gentleman in a restaurant in China with business colleagues who are hosting them (search www.youtube.com

[24]Baxter Magolda, M. and King, P. M., 2004. *Learning Partnerships: Theory and Models of Practice to Educate for Self Authorship*, Stylus, Sterling, VA, p. 5.

[25]Baxter Magolda, M. and King, P. M., 2004. *Learning Partnerships: Theory and Models of Practice to Educate for Self Authorship*, Stylus, Sterling, VA, p. 9–10.

for "HSBC 'Eels' Ad"). The British gentleman finishes his main course of eel and the Chinese colleagues become agitated and order him another bigger eel. The narrator explains "The English believe it is a slur on your host's food if you don't clear your plate. Whereas the Chinese feel that you are questioning your generosity if you do." After clearing his plate a second time even though he has obviously had too much to eat given his peaked appearance, the host orders more again and we see a gigantic eel being wrestled in the kitchen, presumably for yet another massive main course.

ASSIGNMENT 3-10:
INDIVIDUAL ASSIGNMENT – INTERNATIONAL EXPERIENCE INDICATORS SELF-EVALUATION

Complete the *International Experience Indicators* self-evaluation tool on the following pages and see how you score. Use the second and third columns to reassess yourself in a year or two to judge whether you are making progress.

Numbers in brackets are the points to be assigned for any experience. A range in points is indicated to differentiate the extent and/or quality of the experience with regard to how much you believe your international perspectives and understanding were enhanced.

ASSIGNMENT 3-11:
INDIVIDUAL ASSIGNMENT – REFLECTIVE WRITING ON INTERNATIONAL EXPERIENCE

Choose two strategies for increasing your score on the *International Experience Indicators*. Explore how you could implement these strategies and discuss what you would need to do to carry them out.

For example, simpler strategies include reading an international newspaper weekly, learning about the home country of the people in your research group, or inviting a new international student in your program to your home for dinner. High investment strategies include enrolling in a study abroad program, volunteering for Engineers Without Boarders, seeking an international work placement, or taking a language or culture course (Table 3.2).

3.8.1   OTHER RESOURCES ON GLOBAL COMPETENCY

References courtesy of Dr. Laurie Cox, Assistant Dean and Director International Student Services, University of Wisconsin–Madison:

Table 3.2: "International Experience Indicator" rubric (*Continues.*)

International Experience Indicators	Self-Assessment Points (you make the judgment)		
Today's Date:			
Have a passport and have traveled outside the U.S.to:			
• English-speaking country (1-2pts/time with maximum of 6pts)			
• Non-English-speaking country (1-5pts/time with maximum of 10pts)			
Have lived in another country continuously for more than two months:			
• In a city/town of a non-English-speaking country (5-12pts)			
• In a village or rural area of a non-English-speaking country (7-15pts)			
• English-speaking city/town or rural village (3-7pts)			
Expansion of understanding other cultures or global issues from living with or married to someone from another country (1-10pts)			
Work with international colleagues on a regular basis in my work, program, major, or department through my university or outside organizations (1-10pts)			
Interact with and share perspectives about global issues on a regular basis with friends or international colleagues outside the U.S. (1-10pts)			
Currently live or have lived and been active in a neighborhood/community with multi-cultural diversity due to presence of recent immigrants or international people. (1-10pts)			
Hosted international student(s), faculty, scholars, or colleagues in my home			
• For a week or less (2pts)			
• For 1-10 weeks (3-6pts)			
• For more than 10 weeks (7-10pts)			
Had to successfully address a personally embarrassing situation in another culture because of my own cultural ignorance (1-10pts)			
Regularly exposed to international perspectives through reading an international newspaper, news publication or non-disciplinary journal published outside the U.S. and/or regularly listen to international radio or TV broadcasts of news and issues (1-10pts)			
Interact regularly with international people in a club/organization (1-3pts)			
Gave a presentation(s) or lecture in a language that is not my native language (3-7pts)			

Table 3.2: (*Continued.*) "International Experience Indicator" rubric

Knowledge of language(s) other than your native language:						
	Reading	Spoken	Written			
• Language A (1-10pts)						
• Language B (1-10pts)						
• Language C (1-10pts)					.	
• Language D (1-10pts)						
Participated in an education, research, work, or volunteer program abroad						
	In non-English-Speaking Country		In English-Speaking Country			
• Four weeks or less in duration	(4-7pts)		(3-5pts)			
• Four to eight weeks in duration	(7-10pts)		(5-8pts)			
• Semester program	(10-15pts)		(8-11pts)			
• Two or more semesters	(15-20pts)		(11-15pts)			
Completed course(s) that greatly expanded my international competence						
• Course A (1-4pts)						
• Course B (1-4pts)						
• Course C (1-4pts)						
Completed course(s) that somewhat advanced my international competence						
• Course A (1-2pts)						
• Course B (1-2pts)						
• Course C (1-2pts)						
Wrote a research paper on a topic that greatly expanded my international competence						
• Paper A (1-4pts)						
• Paper B (1-4pts)						
Self-evaluation of your openness and understanding of different cultures and your ability to interact with people from different countries (1-10pts)						
Total:						

Althen, G., *American Ways: A Guide for Foreigners in the United States*

Althen, G., *Learning Across Cultures*

Axtell, R., *Gestures: The Do's and Taboos of Body Language*

Axtell, R., *Do's and Taboos around the World*

Chai, M-L. and Chai W., *China A to Z: Everything You Need to Know to Understand Chinese Customs and Culture*

Morrison, T., *Kiss, Bow or Shake Hands*

Nahm, A., *An Introduction to Korean Culture*

3.9   NETWORKING

Your professional network may have more influence on your success that you might imagine. The people in your network can provide valuable feedback on your ideas, technical expertise in areas you are less experienced in, opportunities for collaborations that allow you to approach new research questions, contact with others in your discipline who you would like to work with in the future, and much more.

> **Student Perspective**
> "The most surprising things I learned so far about research would be the importance of professional networking and communication. My previous image of research community is that researchers largely focus on the lab work and have few contacts with anyone besides their colleagues. Therefore, I used to believe academic ability should outweigh any other abilities, and my main focus in the school had always been school work and grades. It was not until I [began a research position] that I realized what I believed was wrong. As I learned more about how research [is] conducted, I found professional networking and communication much more important than I thought."

How do you create a professional network for yourself? You must make connections with people. They can range from casual acquaintances to friendships, but the key is to know other people in your discipline, in other disciplines, and in the community at large. Get to know them and let them get to know you. Offer your knowledge and expertise, when they need it, and they will be likely to return the favor at some other time. These connections can develop within the research group through day-to-day contact, within the program or department through hallway conversation and interactions at seminars and social functions. Connections can be made at the bus, at the gym, or in a coffee shop. The key is to talk to people. If you sit down in class or at a seminar five minutes before it starts, take the opportunity to introduce yourself to the person next to you. Ask them about themselves and engage in a conversation. For some of us this is easier said than done, but with a little practice it becomes more comfortable.

As you become further engaged in your discipline you will begin to attend seminars, workshops, and conferences associated with the topic area of your research. This is an important part of your professional development in several respects: you will have an opportunity to learn about the most recent developments in your field; you will have opportunities to present your own research in either a poster session or a presentation; and you will have an opportunity to broaden your professional network. This last piece is often overlooked, but very important. In order to develop your network, you will need to engage with people informally. This can be done with the people you sit next to before or after a session, joining a hallway conversation or a coffee break, asking someone to have lunch or dinner with you, taking part in a student mixer, and attending luncheons, receptions, or other organized social events. I suggest that students attend these events with a goal in mind. Maybe it is as simple as deciding that you will try to talk with five people you have not met before, or that you will seek out advice about a particular aspect of your research with people that you meet, or that you will inquire about graduate school or postdoctoral research opportunities at their institution. Giving yourself such a goal can help you to overcome any reluctance that you might have to engage in these settings and provide you with a meaningful task that will help you both in terms of building your network and acquiring information that you need.

With a Little Help from Your Friends

Although I have my own professional network, I have found over the years that the networks that my students develop can be just as valuable to our work. I recall a research project where we were stuck on the interpretation of some data we had obtained. The data was produced by a technique that my group had less expertise with than most, but it was critical to the particular experiment we were conducting. One option was to simply repeat the experiment, but it was still a question as to what that would tell us. Instead, we looked for an expert to talk to first. The graduate student working on the project knew of someone in another lab who had used the technique extensively, so he dropped by to chat with him. This developed into a half-hour conversation with several other lab members in this group over coffee. The graduate student walked away with new ideas as to how to approach the problem with a different technique that would help us to interpret our data. This half-hour conversation turned out to be invaluable to the research and saved us significant time.

Student Perspective

"I had the impression that scientists did most of their work in solitude, with essentially all contact being with a few nearby colleagues such as collaborators, advisors, or lab partners. I understood that the goal of science was to share knowledge, but I felt that this was done purely through publications and lectures. I did not notice the existence of any network beyond this. It is true that a scientist does work alone for much of the data-collecting phase of a research project. However, I learned that a researcher must be involved with a larger community to be successful in the profession. Hence, there was a vast network of connections between researchers that I had not noticed … Networking with people that have similar interests in the department is a clear objective, but I learned that it was also beneficial to connect with people at other campuses all around the world."

The other critical aspect of developing your professional network involves broadening your mentoring relationships beyond that of your research mentor, so that you can get different perspectives and a range of constructive criticism, advice, and/or support. Often people think of the mentor-mentee relationship as an exclusive dyad, but in contemporary terms you are seldom the sole mentee in your mentor's life and, even if you were, you can't expect to get everything you need from one individual. Thus, mentoring should occur on multiple levels with multiple individuals, including your research mentor, your peers in and outside your research group, other faculty and staff, and key individuals in your network. You can think of this as a "constellation" of supporting individuals in a variety of mentor-related roles.

Longitudinal research studying career success has shown, that while the quality of your primary mentor significantly impacts your short-term career outcomes, it is the "composition and quality of an individual's entire constellation of developmental relationships that account for long-run protégé career outcomes.[26]" Having many and varied mentors will give you a broader range of perspectives, a wider reaching network, and more opportunities over the course of your career. This constellation of mentors is not something that you create overnight and it frequently grows out of the network connections that you build. You should consider who is already in your constellation of mentors, and watch for other individuals who you can get good mentoring from.

---

[26]Higgins, M. C. and Thomas, D. A., 2001. Constellations and careers: Toward understanding the effects of multiple developmental relationships, *Journal of Organizational Behavior*, 22, 223–247.

ASSIGNMENT 3-12:
GROUP ACTIVITY – WHO IS IN YOUR NETWORK?

Individually:

> Spend 5 minutes listing the people or groups in your network on a piece of paper.

As a Group:

> Discuss the types of people in each individual's network.

> Is your network actually broader than you initially thought?

> Brainstorm about what actions you can take immediately to broaden your network further.

> What strategies can you use to maintain your network?

ASSIGNMENT 3-13:
INDIVIDUAL ASSIGNMENT – DEVELOPING YOUR PROFESSIONAL NETWORK

Identify your goals related to developing your professional network.

> Do you want to improve your network to facilitate your research? Do you want to develop a network that will help you get into graduate school? Or find a job?

Build a list of contacts:

- Identify relevant people.

- If you don't know individual names, identify the types of people you need in your network and then seek out individuals who are that type.

- Identify professional organizations where you might meet people important for your network.

Develop a strategy to court these people individually.

- Using online social networks like Facebook, LinkedIn, and other online tools can help you reach your networking goals.

- BUT meeting someone who is in your network helps to solidify the relationship. How can you arrange to meet each person face-to-face?

- Don't ask for something at your first contact with someone you have just met. And as the relationship develops, take care not to always ask for something every time you interact with a person.

- Reciprocity is important. Try to figure out a way for you to give something. This is why it is important to build your network before you need it.

What can you do to follow up occasionally with these people?

- Schedule time each week to tend your network.

- Be reasonable with the frequency of contact. For some people who you have gotten to know in more depth, your contact may be quite regular but for others it may be as little as once a year.

What courtesies should you follow?

- Respect the time of others.

- Send a thank-you note when someone has provided you with something you truly appreciate.

- Be prepared and willing to reciprocate.

- Ask permission before you use someone as a reference.

ASSIGNMENT 3-14:
INDIVIDUAL ASSIGNMENT – DEVELOPING YOUR PROFESSIONAL NETWORK

Departments on campus and professional conferences often hold social events such as a mixer or reception. These are great opportunities to broaden your professional network. But, it's often helpful to think about how you will start a conversation before you are actually in the position to do so.

After saying "Hi, my name is…," what comes next? You need a strategy to engage with someone you have just met to learn more about them and let them get to know you. So, try asking questions. Depending on the context you can start with something simple like: what's your major? Or. What brings you to this event/place? You might even ask if they have heard about a recent article you have read or ask about what courses they are taking/teaching this/next semester. If you are at a conference you can ask them what sessions they have been attending or comment on a keynote talk that happened earlier in the day.

Brainstorm three questions you could ask or conversation starters you could use to engage in a conversation with someone you have just met. Consider three different situations: talking to a person at a nearby table in a coffee shop or cafeteria; sitting next to someone five minutes before a class or seminar begins; mingling at a social event associated with a professional function.

Now put it into action. Set some goals for yourself, such as: meet three new people over the next week; get to know two other people in my major; talk to someone more senior than myself who is in my professional area.

ASSIGNMENT 3-15:
INDIVIDUAL PROJECT – STARTING YOUR OWN PEER MENTORING GROUP[27]

Step 1: Identify a common topic of interest for the group, e.g., fellowship proposal writing, journal club in your research area, qualifying exam preparation, dissertator support group, etc.

Step 2: Identify a few peers who you would like to invite to join you in the group. Have a conversation with each of them about their interest in meeting regularly on this topic. Identify the best venue for the meetings and timing for the meetings. Take into account that some members may have other obligations that prevent them from meeting at certain times or on certain days of the week.

Step 3: Set up a text group or listserve with the initial members and send out a formal announcement, e.g., an email might include the following: "Thank you for agreeing to join me in our TOPIC group. I have reserved ROOM/BUILDING for our first meeting on DATE/TIME. At this initial meeting I propose that we develop an agenda for our group and plans for our future meetings over the semester."

Step 4: Develop consensus within the group about the formality of meetings, frequency of meetings, optimal size of the group, and responsibilities of the group members.

Step 5: Grow the group to a sustainable size. This can be accomplished through the networks of the initial group members or talking with a staff member affiliated with your degree program about other individuals they may know of who would be interested in the group.

Step 6: As the "convener" of the group, you will be responsible for sending out reminders for meetings and keeping the momentum of the group going. It is good practice to rotate the "convener" responsibility to a new individual for a group that meets for more than a few months.

---

[27] Adapted from Crone, W. C., 2010. *Survive and Thrive: A Guide for Untenured Faculty*, Morgan & Claypool Publishers.

CHAPTER 4

Building on the Research of Others

4.1 THE LITERATURE

Together the collection of journal publications, conference proceedings, handbooks, monographs, books, and student dissertations/theses is referred to as "the literature" and provides a foundation of knowledge for you and others to build upon.

Your primary exposure to engineering concepts may have been through textbooks up until a certain point in your education. As you pursue more advanced study, and particularly research, you will more regularly gain information from journal articles, along with other sources, such as conference proceedings, technical handbooks, and edited collections (books where each chapter is contributed by different authors). The purpose of journal articles is to provide an open report of findings and new discoveries in a timely manner. These papers will contain details that you will not be able to find anywhere else. Certainly the most recent findings in a particular research area will only be available in journal articles and conference proceedings, but you will also find that journal articles published decades ago may also be critical to you in your research. These are sometimes referred to as seminal papers if they contain the origins of a research idea, completely changed the way a topic was understood, or provide results that are continuing to be foundational to the field.

Ideally, you will want to read and rely only on articles that are published in peer-reviewed, archival journals and conference proceedings. These may be available in both paper and electronic formats, but the key issue is the reliability of the information being presented. The archival nature of journal publications also ensures their longevity and provides a searchable record of findings. Just because someone has published something, does not mean that it is correct. However, you will find that more reliable information can be found in reputable journals that have a rigorous peer review process. The "Impact Factor" of a journal will also give you a guide as to the stature of the journal in its field.[1] Regardless of where something is published and by whom, you must look at all information that you read with a critical eye.

---

[1]The Impact Factor of a specific journal is based on a calculation involving the number of times that the papers in that journal are cited by others. These values vary by field so it may be helpful to look at how the Impact Factors of journals within a field compare to each other. You can find Impact Factor information from a variety of sources, but they trace back to the Journal Citation Reports (JCR) and the information is integrated into Web of Science, as well as other indexing systems.

There are a range of different types of journal articles that you will find in the literature which will vary in length and content. Some journals, and the articles published within, will specialize in various ways—for instance you may find a journal in your field that specializes in instrumentation or experimental methods, another that focus on modeling and computation, etc. Additionally, there are short communications, some of which are specifically designed for rapid publication of new findings, full length articles that present original research in full detail, and review articles that synthesize the state of the art concerning a particular topic. You may find review articles to be very helpful, especially as you are entering into a new area of inquiry. A review article, if done well, will not only summarize the research that has come before, but will also synthesize these results, present challenges and future directions for research, and provide some commentary on future directions in the field. Conference papers/proceedings are also prevalent in some disciplines and in some cases may be the higher profile publication of a disciplinary area.

The literature in a research subject area can be very challenging to navigate initially and it is often a good first step to ask your research mentor to suggest key articles that you should read and indicate which journals are the most relevant to your research. This will give you a good foundation to start building your familiarity with the literature in your research area. Your initial exposure will be challenging. But be assured, as you read more, you will understand more of what you read.

4.2   VALUING WHAT CAME BEFORE YOU

The literature contains a vast amount of information which grows at an ever-increasing pace each year. You may feel that there is so much to read and learn about that it is pointless to even begin. But begin you must. With each paper your knowledge and understanding will grow, along with your ability to discern when you have found an important research contribution that will influence your own work. You will also develop the ability to occasionally disagree with and challenge some of the methods, results, and conclusions that have been previously published.

Scientists and engineers read the literature for a number of important reasons: to learn what others have done so we don't reinvent the wheel; to build upon the prior published work in order to advance our own research; to keep abreast of the recent findings from other research groups; to be able to describe how our own research fits into a broader context of the field; and to distinguish our contributions from the contributions of others.

Early in your research career, your purposes in reading the literature may be a bit different. If you are applying to a graduate program and interested in working with a particular faculty member, it will be important for you to become familiar with their prior research. Certainly you will look at their webpage, but it is also important to look at what journal articles they have published in the last few years to get a better idea of the trajectory of their recent research. If you will be meeting with a faculty member on a campus visit, you may want to choose a journal article that this faculty member authored recently so that you can read it prior to your visit.

Although you don't need to understand everything in the paper, you want to read it carefully enough to have intelligent questions to ask regarding the paper. Keep in mind that this is likely to be a publication about a research project that is complete and no longer active. Although it is a good starting point for a discussion, it may or may not be representative of the research that this faculty member is currently doing. In your conversation you will want to find out the direction of their current and future research projects. This information will be important for you in order to determine if there is a good match between your interests and the direction of this faculty member's research group.

> ### Know What You Don't Know
>
> I am always encouraged when a student I have invited to join my research group asks for a few relevant journal articles to read before their position begins. This usually happens with the best of the undergraduate researchers who will be joining the group for a summer research opportunity and the graduate students who will be joining our research group in the next semester. Asking the question is a positive indicator, but then having the follow though to read the article before they arrive is a good sign that they will be a successful researcher. I never expect that the student will understand everything they are reading at this stage. The outcome I like to see is that they show up on the first day with questions about what they have read and how these articles relate to what they will be doing.

When you have joined a research group—ideally even before you arrive in that group—read the papers that the research group has published recently. Not only will this give you some relevant background on the types of research that the group has undertaken and the techniques they have used, the author list of each paper it will also give you an idea of who has been working together—which students, postdocs, and scientists have contributed to which projects, and which other faculty the research group collaborates with most frequently.

An important area of value to you as you embark in a new research area will be the literature most relevant to the project you will be working on. Some of this may have been published by the research group you are joining, but much of it will likely have been published by other research groups around the world. Your research mentor will be able to help you identify some of the most relevant journal articles that you should become familiar with initially. Read these articles carefully and save them—they may be articles that you will want to go back to and reread after you have begun working on your research project. Pay attention to the author lists and watch for new papers to come out from these same authors—those new articles may also be relevant to your work. A later chapter will go into more detail on author order, but it is usually most relevant to note the first author, last author, and corresponding author(s). Keeping abreast of the relevant literature in your area is an important aspect of your development as a researcher.

Eventually, you will be the one pointing out new articles that have appeared in the literature to your research mentor!

Learning How to Read

Over time, everyone develops their own approach to reading journal articles for both efficiency and getting the most information out of your literature search. My own style is to first focus on the abstract and conclusion to decide whether or not I need to spend more time with the article. Then I usually go to the figures next. If a paper is well done, the figures and their captions give you an outline in a visual format. Then I will begin at the beginning and give the article a quick read. If I find that I am still interested and need more detailed information, then I take a second read in a much more thorough fashion. I will work through equations, delve into the methods, carefully compare the figures to the text, question the assumptions made, ponder the conclusions drawn, and decide what it is that I can take away from this paper that will be useful to me. I also look at the references cited as well as who has cited the paper, so that I can find other relevant journal papers to read. For seminal papers on a topic, I may reread them a third, fourth, or tenth time over the course of years.

Each journal article you read will build your knowledge and make you more skillful at extracting the key concepts and pieces of information that you need. Settling on you own approach will come with experience.

You will also need to keep track of the papers you read and the ideas that have come from them (tools for doing this will be discussed in more detail in the section on Citation Management below). When you eventually put together your own research findings for presentation and publication it will be critical for you to cite the work of others. You will need to show that you are knowledgeable about the literature in your area of research and you will need to give credit for the ideas of others that have contributed to your own work. The best researchers are those who show how their own work builds on and extends the field by acknowledging the work of others.

Student Perspective

"Information is continuously flowing throughout the globe and with the elaborate access we have to it, via Internet and also extensive communication modes; an individual has at his disposal a plethora of work, ideas and information about almost everything. This can be at times dangerous, as we as people forget that though we have access to all this information, we don't have ownership over it. These works or ideas aren't ours; they're someone else's intellectual property."

4.3   READING JOURNAL ARTICLES

Journal articles tend to have a similar structure, although some variations may be found depending on the style requirements of the particular journal. An important aspect of reading journal articles is knowing about all the information that is there for you to find. Common components include: author information; an abstract; an introduction and/or background section; a methods or techniques section; results; discussion; conclusion; acknowledgments; and references/citation information. The examples shown on the following pages (Figures 4.1–4.4) come from two samples articles published in peer reviewed journals. The paper by Gall et al.[2] is a research article containing new experimental results, whereas the paper by Maboudian and Howe[3] is a review article providing the current state of understanding on a topic.

Every journal article begins with a title. There are many styles to creating a title, but they should be descriptive of the content. The title can often give you a hint as to whether or not you want to read the article, but the most succinct and descriptive piece of the article is the abstract. It should give you a good idea of whether or not you want to read further.

The first page of the journal article will also contain author names and affiliations. The affiliations will often give you a hint about the disciplinary background of the authors, given the department or unit that they are affiliated with, as well as the institution where the research was conducted. In collaborative projects you will sometimes see authors listed from multiple institutions. For instance, it may be that experiments were conducted at one institution and the modeling work was conducted in a different research group at another institution.

Sometimes the journal will also provide information about the publication timeline. For instance, when the article was first received and when it was accepted for publication. This information together with the actual publication date gives you a sense of how quick of a turnaround this journal does in their review process, which may be important to you when you consider journals to submit your own work to (Figures 4.1 and 4.3).

You will also find the title of the journal, volume and number of the journal, and page number(s) on the first page. Together with the title and author list, you can compile a complete citation for the article. For example:

Salick, M. R., Napiwocki, B. N., Sha, J., Knight, G. T., Chindhy, S. A., Kamp, T. J., Ashton, R. S., and Crone, W. C., 2014. Micropattern width dependent sarcomere

---

[2]Gall, K., Dunn, M. L., Liu, Y., Labossiere, P., Sehitoglu, H., and Chumlyakov, Y. I., 2002. Micro and macro deformation of single crystal NiTi. *Journal of Engineering Materials and Technology*, 124(2), 238–245.

[3]Maboudian, R. and Howe, R. T., 1997. Critical review: Adhesion in surface micromechanical structures. *Journal of Vacuum Science and Technology B: Microelectronics and Nanometer Structures Processing, Measurement, and Phenomena*, 15(1), 1–20.

Author Names
and Affiliations

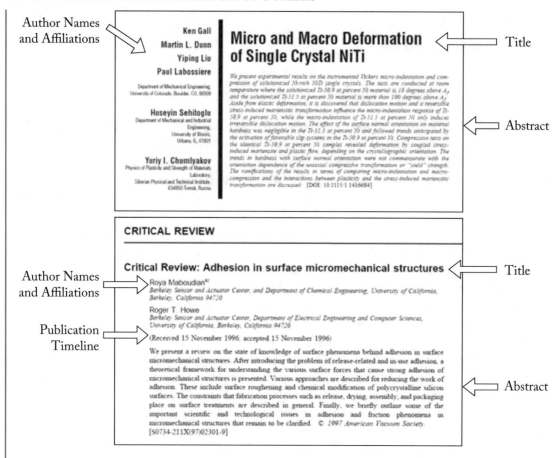

Title

Abstract

Author Names
and Affiliations

Publication
Timeline

Title

Abstract

Figure 4.1: Title, author information, and abstract for a research article (top) and a review article (bottom).

development in human ESC-derived cardiomyocytes. *Biomaterials*, 35(15), 4454–4464.

It should be noted that there are multiple citation styles and most citation management systems will output information in whatever the style you need. The citation shown above uses the Chicago style based on *The Chicago Manual of Style*. Each journal will have a required citation format, some of which adhere to one of the common styles and some that are unique to the journal.

Often on the first page, but sometimes at the end of the article, you will find contact information for the corresponding author (Figure 4.2). This is an individual that you can contact with questions about the paper if you are delving deeply into its content. This contact information should not be used frivolously, but if you have a serious question that you have not been able

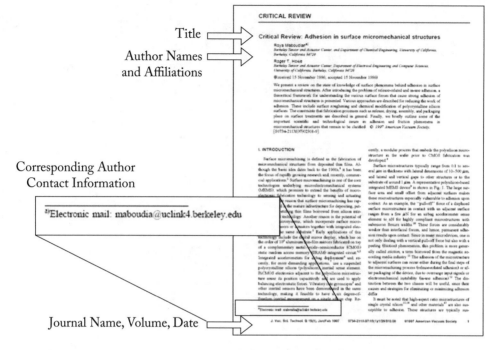

Title

Author Names
and Affiliations

Corresponding Author
Contact Information

Journal Name, Volume, Date

Figure 4.2: First page of a sample review journal article.

to find the answer to in any other way, it is reasonable to make contact with the corresponding author.

Although some short papers will not separate out the article into sections, most will have some variation of the following section titles: Abstract, Introduction, Background, Methods, Results, Discussion, Conclusion, Acknowledgments, References. You will notice throughout, and particularly in the Introduction/Background sections, that prior published work related to the subject has been cited (Figure 4.3). Even if the current article does not contain exactly the information you are looking for, these citations will refer you to other references that may be useful. Following this trail of breadcrumbs can sometimes be more fruitful than a search engine because the work being cited has been read by these authors and deemed useful and relevant enough to include in their article. In a way, this is an additional level of review and thus a reason why these papers are worth taking a look at beyond the many others that may pop up in a literature search on the topic.

Depending on what you are trying to glean from a particular article may mean that you will spend more or less time with it or delve into more detailed reading of particular sections. In many cases, you are simply interested in the findings and what the authors have contributed to the field. The Discussion and Conclusion are likely the sections of the article that you will read more than once. Or, if you are an experimentalist, and you are specifically looking for a

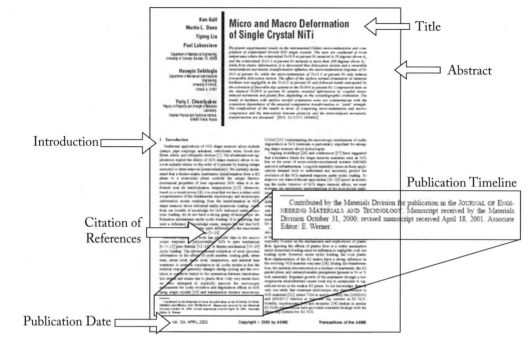

Figure 4.3: First page of a sample research journal article.

method relevant to a process or technique you need to accomplish in the lab, then you may be most interested in the Methods section. In a different situation you could be trying to figure out the best way to show the data that you have collected or produced, so the figures in the Results section would be your focus. Keep track of the papers you have read—ideally using a citation management system like what is discussed later in this chapter—it is likely you may want to cite the paper in something that you will write later (e.g., a report, thesis, or journal article of your own) and you may want to reread that paper again at a later date. It is often the case that you get more from a article on a second (or tenth) reading, especially when you have had a chance to learn more about the topic through your own research and other things you have read in the literature.

Usually at the end of the paper you will find an Acknowledgments section. This might include individuals or facilities that aided the research, but whose contributions did not rise to the level of authorship. It will also include the funding source(s) that made this research possible (Figure 4.4). In some cases this might indicate a possible conflict of interest, or bias, in the interpretations of the findings because of this support (for instance, there have been accusations of researchers supported by Google who wrote about Google in articles but did not acknowledge that they had received funding from them). The funding sources listed will also give you an idea of what federal agencies or foundations are interested in supporting this type of work. It is

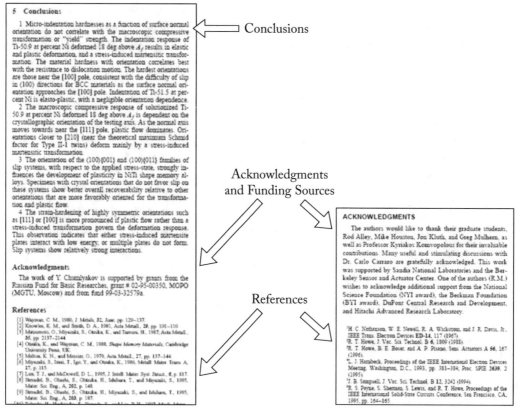

Figure 4.4: Conclusions, Acknowledgment, and References sections of two sample journal articles.

possible that you may want to look to apply for funding from one of these sources in the future, through fellowship or grant opportunities.

The References section gives you the complete information for all of the citations in the paper (Figure 4.4). This is a valuable resource for you because the paper has identified other relevant literature which you may be interested in reading.

Finally, the last item to be aware of is Supplemental Material. A growing expectation in publication is for the authors/journal to provide additional on-line supplemental content that is relevant to the article. Usually a link will be given somewhere in the article that sends you to a page with additional content on the publisher's website. You might find in the supplement that more results are available, details of the methods are given, or code used in the research is being provided for others to download and implement. Sometimes critical information for your research will be found in the Supplemental Material, so do not forget to watch for a link to it!

Remember that for an important article with high relevance to your research, you should expect that you will need to read it more than once. To really understand what you need to, you may also need to do some additional learning outside of the article and talk to others about the meaning of certain aspects.

> **Student Perspective**
>
> "Learning how to effectively locate relevant papers and navigate through scientific literature is a critical skill to develop. Simply finding the right journal articles is only part of the process of conducting a thorough literature search. In order to actually gain understanding and benefit from scientific publications, I need to develop skills in critically analyzing the research methods and conclusions which are presented. This skill is developed with practice in reading journal articles, which will help with gaining familiarity with projects related to mine and their associated terminology. As my research advisor has pointed out, fully comprehending the meaning of a certain publication often requires more than one reading, even for someone well versed in the subject matter. It is therefore best to be honest with myself about what I do and do not understand, and to give myself time to become knowledgeable."

ASSIGNMENT 4-1:
INDIVIDUAL ASSIGNMENT – SUMMARIZING WHAT YOU HAVE READ

Choose a journal article of interest to you. Read the article and become familiar with the main points being put forward by the author(s). Summarize the article in a short paragraph, highlighting the main points that you have identified. Refrain from just rewording the abstract that was written by the author(s). Write from your own understanding of the article, even if you feel that understanding is incomplete. Use your own words, even if they are not as technical as the one used in the article.

ASSIGNMENT 4-2:
GROUP ACTIVITY – JOURNAL CLUB

A strategy for becoming familiar with and keeping up with the literature is what is commonly referred to as a Journal Club. These are used more often in some fields than in others. Some research groups hold their own journal clubs with faculty participating. Some groups of graduate students take it upon themselves to create a Journal Club group in order to help each other read, understand, and interpret the literature.

Journal Clubs run in a variety of ways, but they have commons features: the group has a research theme, everyone participates by choosing and reviewing journal articles as well as commenting on the ones chosen by others, and the end goal is for everyone to increase their knowledge of the topic area. In many cases the expectation is that everyone has looked at the article prior to the Journal Club meeting.

K. Barker suggests the following guidance for discussing a paper in Journal Club fashion in her book *At the Helm*[4]:

- "Summarize the main point of the paper."

- "Describe the paper in detail."

- "Analyze the data."

- "Itemize the strengths and flaws in the paper."

- "Compare the paper to other papers."

- "Is the paper well written and the data clearly presented?"

- "Predict the next step in the research."

As a Journal Club presenter you should come well prepared. This is important so that everyone is making the best use of their time, and it is part of showing that you are a professional who takes research seriously. Your ability to present a paper successfully will improve over time, as will the depth of analysis you can bring to each paper you review. As you begin in this process, choose a paper that is central to the research you are conducting.

When my students present in lab meeting about a paper they have read, I suggest that they attempt to answer the following questions in their presentation of the work.

- Describe the paper.

    - Who authored this journal paper? What institution are they from?
    - How is that researcher or group related to your research group?
    - What is the problem being studied? How is this problem related to your research?

- Summarize the main point of the paper.

    - What are the key methods used?
    - What are the main results?

---

[4]Barker, K., 2002. *At The Helm: A Laboratory Navigator.*

- Detail both the strengths and flaws of the paper.

  - Is the paper well written and the data clearly presented?
  - What are the assumptions in the paper? How realistic are they?

- Take a deep look into the data presented.

  - What story does each figure tell?
  - Is there supplemental data provided that should also be considered?
  - How sensitive are the results to the assumptions?
  - What did you learn from this paper? How is this relevant to your research?

- Compare the paper to prior published work and relate it to your own research.

  - What are the similarities and differences of this research compared to other related research?
  - How does it connect with your research?
    Similar/different approach/methods/findings?

- Discuss potential opportunities for future work based on this paper's findings.

  - What do you think the authors are working on now?
  - What would be a natural extension of this work?

Note that you may not be able to address all of these topics given the time constraints, so you may have to prioritize.

4.4   READING CRITICALLY

It's important to note, especially for those new to research, that not everything that is published is perfect, or even correct. There are a variety of reasons for this. Some of them quite innocent: maybe the understanding the field has changed/deepened since the article was written so it is no longer the appropriate methods or interpretation. Maybe there was an error in the publication process that made an equation incorrect (check to see if there is an Errata for the paper because the error may already have been discovered). Maybe the results are reported accurately, but the interpretation might be made differently by others. Or, maybe the paper is just poorly written and difficult to read.

Student Perspective
"Before this semester, I generally interpreted published research as always having the most accurate information about a subject. However, I

learned in class and from my research mentor that sometimes journal articles contain inaccuracies. My most vivid example of this was when my research mentor asked me to read an article relating to the project I was interested in. After reading the article, a theory mentioned still was unclear and I could not find any background information online. When I asked my mentor about the theory, he said that the group that published the research was more interested in producing a product than explaining a phenomena and the theory they proposed was not very sound. Indeed, it seems that some published articles offer the chance for the scientific community to debate and arrive at a conclusion rather than accept an article as fact."

Unfortunately, there is also the darker side of error, negligence, and misconduct where what has been published was intentionally misleading or incorrect. This may be the fault of one, some, or all of the authors. This topic will be discussed in more detail in Chapter 5.

Our role as a researchers is to always question, rather than take things at face value. This applies to your own results and conclusion in addition to the results and conclusions of others. As you read about the work published by others, you are determining if the research is trustworthy and done in a reliable way by examining the methods used and determining if the conclusions are supported by evidence. One way to do this is to draw your own conclusions about the work before reading the conclusions section and then compare yours to those of the authors. This will be very challenging to do initially, but as you gain more experience it will become easier.

Among the recommendations for analyzing an engineering document in the advice of Paul, Niewoehner, and Elder[5] are the following items that focus on reading critically:

Data

"Is the data accurate? How was accuracy established?"

"Is there data missing? Is there adequate data?"

"Is the data of sufficient quality?"

"What controls were applied to isolate causal factors?"

"Is the entire dataset presented? What criteria were used to select the presented data sample from the complete set?"

Concepts

"Are the appropriate theories applied?"

---

[5] Paul, R., Niewoehner, R., and Elder, L., 2007. *The Thinker's Guide to Engineering Reasoning*, The Foundation for Critical Thinking. www.criticalthinking.org.

"Have alternatives concepts been considered?"

"Are concepts used justifiable?"

Point of View

"Are there competing theories that could explain the data?"

"Have alternative ways of looking at the situation been avoided in order to maintain a particular view?"

Assumptions

"Are the assumptions articulated/acknowledged?"

"Are the assumptions legitimate or necessary?"

"Do the assumptions take into account the problem's complexity?"

Conclusions

"Are there alternative conclusions?"

"Is speculation misrepresented as fact?"

"Do the conclusions follow from the assumptions?"

"Is further testing required?"

Last, you should determine if the authors or others have found problems with the papers you are citing. Journal articles can be retracted after they are published. John M. Budd, a professor in the School of Information Science and Learning Technologies at the University of Missouri at Columbia, reported that between 1999 and 2009, 1,164 articles were retracted from biomedical journals and "55% of the articles included in this analysis were retracted for some type of scientific misconduct.[6]" Jennifer Howard, a Chronicle of Higher Education Reporter, advises: "Authors, you really ought to take a look at the journal articles you cite. Not only is it the responsible thing to do, it will save you the embarrassment of discovering after the fact that you have given a nod to a retracted or discredited paper.[7]" If you are not downloading and looking at the article yourself because you are simply copying someone else's citation you will not see the retraction notices. You need to get a copy of the original source, read it, and draw your own conclusions in order to cite the work reliably.

[6]Budd, J. M., Zach C. C., and Anderson, K. M., 2011. Retracted publications in biomedicine: Cause for concern. In *Association of College and Research Libraries Conference*, pp. 390–395.

[7]Howard, J., 2011. Despite warnings, biomedical scholars cite hundreds of retracted papers. The Chronicle of Higher Education.

4.5   LITERATURE SEARCH

A literature search begins with a topic and the key terms that are relevant to that topic. An important first step is to get familiar with the terminology and jargon that is relevant to your research. Attend seminars, engage in research group meetings, and talk with others in your research area to begin becoming fluent with the terminology and jargon commonly used. Wikipedia can be a useful resource to get a basic understanding of the topic and terms. Your research mentor may also suggest some textbooks or journal articles that are helpful.

At some point you may be specifically asked to conduct a literature search. Your research mentor may give specific key words to use, a question to answer, or a general topic area for you to explore. Regardless of whether you are specifically requested to do so or not, conducting literature searches is an important part of the research process that you will need to undertake. In the early stages of your research project you may simply be looking for background information, and a general idea of what has already been published on your topic area. Later you may need to go to the literature to answer a specific question, such as the details of an experimental technique of an algorithm. Literature in your specific research field is important, but also keep in mind that literature in related fields may also provide you with valuable transferable information. As you continue on your research project, you will also need to keep up on the most recent publications periodically so that you can be certain that your work has not been "scooped" by someone else. If someone publishes specifically on your research topic, you will want to know about it right away so that you can work with your research mentor to determine if you still have complementary research to publish in the area or if you will need to adjust the focus of your research so that you can produce results that distinguish you form the previously published work in the field.

Check with other individuals in your research group and the librarians at your institution to determine what resources are available to you, both in terms of abstract and indexing databases where you can search for relevant content and access to electronic copies of the articles you are interested in. For instance, Web of Science is one of my favorites, but you may find that there is a different database that is more relevant to the literature search that you need to do in your research area. Other abstract and indexing databases are freely available like Google Scholar and PubMed. Some journal articles are available free of charge as open access. However, many others that you would otherwise have to pay for will be available through your institutions library, either directly or through an interlibrary loan function. You will need to talk with others in your research group and the librarians at your institution to determine what is available to you.

> **Finding that Golden Nugget**
> Panning for gold is a good analogy for the literature search; you are looking for the gold nuggets hidden among a lot of sand and gravel that that you will need to sift through. One of the challenges is distinguishing between useful and irrelevant papers so that you can identify information that

addresses your needs. This can be harder at first, but once you have built up some experience you will develop skills that make it quicker. I have always personally enjoyed the process of the literature search; I love the accomplishment of finding that gold nugget, and using it to help me bring about the research idea that I have in mind.

Student Perspective

"And even though looking back it took me a laughably long time to complete this initial literature review, it definitely taught [me] an important lesson. Of course I learned how to more successfully carry out a literature review; which search engines to use, how to intelligently read papers and look into other references etc. More than that though, I learned how to break a task up into manageable chunks; this makes the task seem much less daunting, which personally gives me much more confidence and optimism."

There are numerous strategies for approaching the literature search. They fit into two basic categories: start broad and sift down vs. start narrow and expand up. Each can be employed successfully and often you will want to use a combination of strategies, but let's consider one example of how you might begin. When I am just entering into a new area, I like to start broad, so that I get a sense of the scope of the literature that is available. Once you include all of the relevant terms to your search you will likely have a mountain of literature to look at. When employing this strategy, I will need to quickly eliminate items that have come up in my search by deciding the things that I am NOT interested in.

Take functionally graded materials for instance. In Google Scholar I might start by simply typing those three words: functionally graded materials. In Web of Science I might search on the topic using: functional* grad* material* (the * is a wildcard symbol that tells the database to include things that have alternate endings like functional and functionally, graded and gradient, material and materials). In Web of Science this search gives me nearly 20,000 results. Yikes! It's a HUGE research area, but it is unlikely that all of these results are relevant to my interest. Maybe I am most interested in polymeric materials, so I refine my search by using polymer* as another key word with the AND operator requiring both terms to appear in the database record. This has already narrowed it down to a little over 1,000. The research application I am interested in also requires that the material is biocompatibility. I'm not interested in the biocompatibility research itself right now, but I want to make sure the materials can be used in the way I want, so in that case I might want to include a term like cell culture in order to include only those materials that are useful in that circumstance. This narrows it down to a more reasonable 45 results. Now I'll start sorting through titles and abstracts to decide which ones of these may be relevant to my research. I still have a lot of papers to consider but I'd at least narrowed down my search by including a few additional parameters.

When I am looking at the results that a search engine produces for me, I want to consider the titles first. For the ones that look relevant I will look through the abstract and excluded many of the ones I read, because this added information tells me right away that they're not likely to be very useful to me. When I do find an abstract that looks relevant, I can then open the PDF of the article and skim through the paper. I personally tend to focus on the figures first, and if some of them look relevant read the conclusions next to see if there is likely to be useful information in the article. At this stage I don't necessarily want to read the whole paper because I'm still in the middle of literature search, so if the abstract and the conclusions look promising I will save that PDF and come back to it later for more detailed reading. Even of the ones I have saved, only a fraction will ultimately be useful to me when I read them more thoroughly.

As you are doing the search it is always important to also think about different ways in which the terminology may be used. After you have read through a number of titles and abstracts, you may realize that there is an additional term or refinement of a term that may give you more relevant results. You are seldom done after just one try and you will probably want to re-run your search with slightly different key words after you have become more familiar with the topic. It's also important to realize that not all databases have access to the same collections of resources. You may need to use more than one. Often your research mentor, or a reference librarian, can point you toward the most relevant databases for your discipline. A librarian can also tell you how to access search engines and databases remotely, so that you can access the literature even when you are off campus.

Once you find some papers that are particularly relevant to your research question, you can use those with the search engines to look both forward and backward in time. Start with an article that you had already found to be relevant, then go backward and forward in time by looking at the papers that this article cites and who have cited this article since it was published. Abstract and indexing databases like Web of Science and Google Scholar have this functionality. You might find some nice new nuggets this way. You can also set up a various different alters with most of these systems to notify you of future publications when they appear. If you have developed a set of useful search parameters, you can set an alert to notify you whenever an newly published article meets those same parameters. A citation alert on particularly relevant article that you have already identified will notify when another new article comes out that cities this previous one. You can also set up a citation alert on key authors in the field so that you are notified when they have a new publication appear. However, even with these alerts in place, you will need to regularly revisit your literature search to identify if new publications have appeared.

Review articles can also be an excellent place to resource if one has been written in your research area. Happily, the authors of the review article will have already done the hard work of pulling together and summarizing the relevant research papers on the topic. However, your search does not end with the review article. You will want to get a copy of the most relevant papers that the review article cites and read them yourself. You will also want to see what other papers have cited the review since it has been published because there may be more recent papers that are also important for your research. To find review articles, you will want to use the key words relevant to your topic and include the word "review" if you are searching in an index like Google Scholar, or in a database like Web of Science, there is a Review button you can click. This will narrow your search results to only the ones that have been identified as review articles, so that you can take a look at those in more detail.

You also need to recognize that journal articles may not be the only source of relevant information on your topic. You will want to look at patents, government reports, handbooks, and books that are available through the internet, or your institution's library. Some of these resources can be a little bit more difficult to identify. I have found over my career that the reference librarians can be an incredibly valuable resource for helping you to identify how to access the information that you need. If you have such a person available at your institution, you should take the opportunity to get to know them.

ASSIGNMENT 4-3:
INDIVIDUAL ASSIGNMENT – LITERATURE SEARCH WITH KEY WORDS

Work with your research mentor to identity key words for a useful literature search that you can undertake and get their suggestion on which databases to use in your search. Conduct the literature search using these key words and identify the 5–10 most relevant journal articles on this topic.

ASSIGNMENT 4-4:
INDIVIDUAL ASSIGNMENT – PATENT SEARCH

Use the United States Patent and Trademark Office website at `www.uspto.gov` to conduct a patent search using key words that are relevant to your area of research. (Note that this website provides a number of resources for how to conduct a search effectively.) Identify 1–2 patents most closely associated with some aspect of your research and summarize your findings.

ASSIGNMENT 4-5:
INDIVIDUAL ASSIGNMENT – GOVERNMENT REPORT SEARCH

Conduct a search for technical government reports that are relevant to your area of research. You may need to work with your reference librarian to identify a relevant database, or you may be able to start with one of the following:

U.S. Department of Defense, Defense Technical Information Center, `www.dtic.mil`

NASA Technical Reports Server, `ntrs.nasa.gov`

U.S. Department of Energy Office of Scientific and Technical Information, `www.osti.gov`

ASSIGNMENT 4-6:
INDIVIDUAL ASSIGNMENT – DISSERTATION SEARCH

Students completing a master's degree often do so with a research thesis, and all Ph.D. students produce a dissertation on their research as the main degree requirement. These are often archived documents and many of repository such as Proquest. Identify five relevant thesis or dissertation titles by conducting a search through your university's library catalog/database or by using Proquest Dissertation Express at `https://dissexpress.proquest.com/search.html`. (If you find that you are interested in learning more about one of these titles, contact your library about access before ordering a copy. Often you can obtain access to a copy through your institution's library.)

ASSIGNMENT 4-7:
INDIVIDUAL ASSIGNMENT – ALTERNATIVE SOURCES SEARCH

Use your institution's library to identity e-books and online handbooks that are relevant to your area of research. Identify 3–5 relevant resources. Choose one of these and write a one paragraph summary of the resource and how it is related to your research interests.

4.6   PROPER CITATION

When you discuss someone's work, either in a presentation or in writing, you need to identify where those ideas came from originally. By including the citation, you have identified to the listener or reader that what you are discussing has its origins in the work of others. The format of the citation may vary depending on the requirements of who you are preparing the work for (e.g., your instructor or a journal). The easiest way to include a citation in written work is to use a footnote (the reference is included at the bottom of the page) or an endnote (the reference is included at the end of the paper).

In the case of a presentation where you have included an image or figure from a source such as a journal article or a website, it is most common to provide the reference information directly on the slide. In this case a short citation is fine, but you will need to provide enough information so that someone else can find the original source. For instance, you might use one of the following formats:

Mature primary cardiomyocyte image from `www.e-heart.org`

Image Credit: Srivinasan, Protocol Exchange, 2011.

[Salick, M. R., et al., *Biomaterials*, 2014]

Even if the material is unpublished, you need to provide credit to the source. For instance, you might credit a colleague with: J. Rogers, with permission. Or for your own work not yet published: B. N. Napiwocki, et al., in preparation.

For written work, there are a number of ways in which the citation may appear. As you are reading journal papers you will have noticed superscript numbers, numbers in brackets, or parenthetical notations that include author names: superscript[1], some type of parentheses (1), the name of the author (Crone, 2010) or authors (De, et al. 2002). These citations identify concepts and results that are culled from other sources and the notations refer you to the particular source in the References section at the end of the article.

For instance, the following sentence appears in the Johnson, et al., 2004 article whose citation is given above:

"In an aqueous environment, stimuli-responsive hydrogels undergo a reversible phase transformation that results in dramatic volumetric swelling and shrinking upon exposure and removal of a stimulus. Thus, these materials can be used as muscle like actuators [1], fluid pumps [2], and valves [3]."

The numbers in brackets, e.g., [3], refer to the reference section of the paper. In this case, an article detailing the use of a stimuli-responsive hydrogel for each component application is given. If you were particularly interested in valves, then you would want to look at reference 3.

In other journals, instead of the number appearing in brackets [ ] or parenthesize ( ), it will appear as a superscript as in this example[8]:

"One of the drawbacks of coil embolization is coil compaction over time, leading to recanalization of the aneurysm. Some degree of aneurysm recanalization occurs in as many as 20% of cases.[1-3] In larger aneurysms, placement of multiple coils can be time consuming, and longer procedural times may lead to increased morbidity and mortality.[3] An alternative to coil embolization is the use of liquid embolic agents. It is thought that filling aneurysms with such polymers will reduce many of the shortcomings associated with coiling, such as coil compaction.[4,5]"

The other common style you will see includes the author and year of the publication in the body of the text. Although this makes the sentence longer, the reader does not have to repeatedly look back to the references section to see whose work is being referred to. For example[9]:

"Previous studies have reported lineage reprogramming into a diverse range of differentiated cells types, including neurons (Vierbuchen et al., 2010), hepatocytes (Sekiya and Suzuki, 2011), and cardiomyocytes (CMs) (Ieda et al., 2010; Song et al., 2012)."

The abbreviation "et al." will often appear in citations. This refers to the Latin phrase *et alia* meaning "and others." When the author list is long, usually more than two, the first author's name is given and followed by et al. to indicate that the paper had multiple authors although all of their names are not listed. Generally, in the references section, the entire author list is included. Every citation style is a bit different and some journals will even deviate from the more common citation styles (e.g., Chicago, IEEE, APA), so you will have to adapt your own writing depending on the requirements.

4.7 CITATION MANAGEMENT

If you are not already familiar with one, now is the perfect opportunity to learn about citation management systems. Keeping track of all the journal articles and other references that you will begin to accumulate on your research project will quickly turn into a big organizational challenge. Happily software systems have been developed that can be a huge timesaver for you, allowing you to easily collect relevant references, organize them, and cite them in your writing.

Examples if such programs include EndNote, Zotero, and RefWorks, but there are many to choose from. Your institution may provide you access to one of these programs, or you may decide to purchase one of these pieces of software for yourself. If you are new to a research group,

---

[8]Moftakhar, R., Xu, F., Aagaard-Kienitz, B., Consigny, D. W., Grinde, J. R., Hart, K., Flanagan, C. E., Crone, W. C., and Masters, K. S., 2015. Preliminary in vivo evaluation of a novel intrasaccular cerebral aneurysm occlusion device. *Journal of Neurointerventional Surgery*, 7(8), 584–590.

[9]Lalit, P. A., Salick, M. R., Nelson, D. O., Squirrell, J. M., Shafer, C. M., Patel, N. G., Saeed, I., et al., 2016. Lineage reprogramming of fibroblasts into proliferative induced cardiac progenitor cells by defined factors. *Cell Stem cell*, 18(3), 354–367.

you should ask your research mentor, or the group members, if the group has a designated citation management system. In some research groups, citation management system content is shared among group members. Although the different citation management systems all contain the same basic functionality, each is a bit different and it may be helpful to look into the details of the options available or talk to a reference librarian before you choose one for yourself.

> **Spend Time to Save Time**
> I was slower than I should have been to adopt a citation management system. When I finally did so, I realized that I had wasted enormous amounts of time by not doing it sooner. My suggestion to you is to start early and save yourself the time upfront! Even so, moving into a citation management system after you already have a collection of journal articles is not as overwhelming as it might seem and fully worth the time investment.

The basic functionality of these systems comes into play beginning with your literature search as you identify articles that are particularly important to your research. Many of the databases that you will use (like Google Scholar and Web if Science) have the functionality to automatically insert the citation into your management system with the click of a button. In most cases, the citation management systems can be connected with your institution's library so that the system can pull the PDF of the article from the library into the management system. Additionally, you can use the management system to organize these references and make additional notations and comments about them as you read and utilize them further. Finally, when it comes time to cite the article as a reference in a paper or report that you are writing, many of the word processing programs available have add-ons that work with the citation management system so that you can easily integrate your citations without a lot of extra work. If you have ever built and formatted a bibliography by hand, you know what a time consuming and irritating task that can be. With one of these systems fully in place, the insertion of a citation into your paper is just a few clicks of a button; the citation is attached to the appropriate place in the paper, and the bibliographic information is included at the end of the paper in your references section.

These systems also allow you to make separate project folders so that you can keep related references bundled together and easier to find. This may not seem very important when you are just starting out, but it will be very handy to have your citations grouped as the number of them grows over time. Using folders or groups also means that you can easily use your citation management system, not only for your research, but also for your coursework and other projects that you undertake.

ASSIGNMENT 4-8:
INDIVIDUAL ASSIGNMENT – INVESTIGATING HOW CITATION MANAGEMENT SYSTEMS WORK

Find a peer or research group member who uses a citation management system. Talk with them about how they use the system and its functionality. Identify at least two new functions or tips about usage that you did not previously know about.

ASSIGNMENT 4-9:
INDIVIDUAL ASSIGNMENT – COMPARING CITATION MANAGEMENT SYSTEMS

Choose two citation management systems and compare their functionality. Describe the pros and cons of the systems in a two-paragraph summary. Consider at least six of the following topics in your comparison:

- Cost (short term while you are a student and long term after you have left the university)

- Operating systems requirements

- Plugins available for word processing programs such as Word and LaTex

- Attachment limits for article PDFs

- Ability to annotate with your own notes and PDF markups

- Ability to create new or edit existing citation styles

- Folder organization and sorting capabilities

- Duplicate citation detection

- Capability to collaborate and share with others in your research group

- Export options between other citation managers

4.8   PREPARING A REVIEW

There are a variety of instances in which you may be asked to review the written work of others. As a student, this most frequently occurs in course settings where you may be asked to do a peer review on a paper written by another student in the course. Alternatively, an instructor may assign a journal paper review as an assignment in an advanced engineering course. More advanced engineering graduate students may even be asked to provide input on a review of a manuscript submitted to a journal. This might be done in collaboration with your advisor or through your advisor's recommendation. Regardless of the particulars of the situation, there are numerous commonalities to the review process. A later chapter will focus on providing feedback to a colleague in a classroom setting, such as a writing workshop. Here we will focus on providing a critical review of a journal article.

> **Get Critical**
>   The first time I was asked to review a journal article, it was for an assignment in a first-year graduate course. We were told to choose an article of interest and turn in a critical review. That was the extent of the instruction and I had no idea of where to start!
>   In many ways, this section is written for those of you faced with such an assignment. However, even if you don't have to do a critical review for a course, it is a good habit to always read critically and this information should help you get started on the path to doing so.

It is important to first understand the expectations of those asking for the review. In the case of a journal, they are very likely to provide you with a set of criteria, or some brief instructions, on the feedback that they would like to receive. In addition to looking for a good article, they want to make sure that the article is a good fit for their journal. You will also have access to the journal's scope through its website. You can use the scope information to tell you whether the manuscript fits with the journal to which it has been submitted.

You will need to begin by reading the manuscript thoroughly. It will likely require multiple passes through the manuscript in order for you to complete your review, but in the first reading you will get a general sense of the article. You may also take away an impression of its overall strengths and weaknesses in this first reading. During this first reading keep in mind a few questions.

What is the key takeaway message?

Does the abstract give a compelling and yet reasonable summary?

What points did you find initially confusing?

Are the figures clearly presented?

Are the terms defined and the equations understandable?

Do the conclusions follow from the results?

What is the significance of the findings presented?

Is prior published work appropriately cited?

Is the paper written with good organization and grammar?

In your review you will be seeking to help the authors improve their manuscript so that the future readers can easily understand it. You will need to be on the lookout for both scientific problems in the methods and analysis, as well as writing issues such as clarity and presentation. Your review should be detailed enough to help the authors improve their manuscript regardless of whether or not you recommend to the editor that it be accepted for publication in this particular journal. The decision of whether or not to publish will ultimately be the editors to make, but you will need to give an option if it should be accepted (with minor or major revisions) or rejected based on the quality and impact on the field.

As you get more involved in your research area you will begin to learn which journals are the most important in your field and you should become familiar with their scope and criteria for publication as you begin to work toward publishing your own research. Every journal has defined its scope to identify what research it will publish. For example, the journal *Experimental Mechanics* is published by Springer with the Society for Experimental Mechanics, a professional organization that I have been a member of for 30 years. If you go to the journal website you will find the following information describing the scope of that journal[10]:

- Explores experimental mechanics, including its theoretical and computational analysis.

- Addresses research in design and implementation of novel or enhanced experiments to characterize materials, structures, and systems.

- Spans research in solid and fluid mechanics to fields at the intersection of disciplines such as physics, chemistry, and biology.

- Extends the frontiers of experimental mechanics at both large and small scales.

Below are some example criteria provided to the reviewers to give you an idea of what is commonly requested. Each journal will give instructions to its reviewers and may have special-ized criteria to consider, but the recommendations for reviewing provided by Springer Interna-tional Publishing are representative of the type of requests you would see.

Evaluating Manuscripts[11]

When you first receive the manuscript it is recommended that you read it through once and focus on the wider context of the research.

Springer Publishing recommends that you ask questions such as the following.

---

[10]*Experimental Mechanics*, Springer International Publishing, `https://www.springer.com/engineering/mechanics/journal/11340`.

[11]*Evaluation Manuscripts*, Springer International Publishing, `https://www.springer.com/us/authors-editors/authorandreviewertutorials/howtopeerreview/evaluating-manuscripts/10286398`.

- What research question(s) do the authors address? Do they make a good argument for why a question is important?

- What methods do the authors use to answer the question? Are the methods the most current available or is there a newer more powerful method available? Does their overall strategy seem like a good one, or are there major problems with their methods? Are there other experiments that would greatly improve the quality of the manuscript? If so, are they necessary to make the work publishable? Would any different data help confirm the presented results and strengthen the paper?

- Were the results analyzed and interpreted correctly? Does the evidence support the authors' conclusions?

- Will the results advance your field in some way? If so, how much? Does the importance of the advance match the standards of the journal?

- Will other researchers be interested in reading the study? If so, what types of researchers? Do they match the journal's audience? Is there an alternative readership that the paper would be more suitable for? For example, a study about renal disease in children might be suitable for either a pediatrics-centric journal or one that is targeted at nephrologists.

- Does the manuscript fit together well? Does it clearly describe what was done, why it was done, and what the results mean?

- Is the manuscript written well and easy to read? If the manuscript has many mistakes, you can suggest that the authors have it checked by a native English speaker. If the language quality is so poor that it is difficult to understand, you can ask that the manuscript be corrected before you review it.

After your first reading, write one or two paragraphs summarizing what the manuscript is about and how it adds to current knowledge in your field. Mention the strengths of the manuscript, but also any problems that make you believe it should not be published, or that would need to be corrected to make it publishable. These summary paragraphs are the start of your review, and they will demonstrate to the editor and authors that you have read the manuscript carefully. They will also help the editor, who may not be a specialist in this particular topic, understand the wider context of the research. Finally, these paragraphs will highlight the manuscript's main messages that will be taken away by readers.

You can then proceed in evaluating the individual sections of the paper. (Note that Springer's website gives additional detailed questions to consider in each section of the manuscript.)

Most engineering journals use a "closed" peer review process where you will know the identities of the authors, but they will not be informed of your identity. Even though your review will be anonymous in this sense, you should always behave respectfully and professionally in your review. It is also likely that your criticism will be better received if you note what the authors did well, in addition to what they need to improve. Additionally, keep in mind that as a reviewer you are seeing research before it is published and publicly available, but you must keep this information confidential until the publication is released.

As a future author of a journal paper the review criteria described above and the practice of being a reviewer will be helpful in allowing you to evaluate your own manuscript with a critical eye prior to its submission.

ASSIGNMENT 4-10:
INDIVIDUAL ASSIGNMENT – WRITING A JOURNAL ARTICLE REVIEW

Choose a journal article relevant to your area of research. Conduct a 2-page written review. Begin with a 1–2 paragraph summary of the paper. Conduct the remainder of the review using the "Evaluating Manuscripts" guidance from Springer above or the review criteria from the journal in your area of research (find the journal's website and look for the guidelines for referees/reviewers).

ASSIGNMENT 4-11:
INDIVIDUAL ASSIGNMENT – ANNOTATED BIBLIOGRAPHY

Compile an annotated bibliography of a research topic of your choice. This topic may be related to a seminar that you have attended this or a research topic in your own subdiscipline area of interest.

Your annotated bibliography must contain a minimum of eight journal articles. For each article you must give the full citation (using a standard style such as APA or Chicago) and a brief description (roughly 150 words) of the main purpose and findings of the paper. Include a topic title at the top of the top of the bibliography.

4.9   CREDITING THE WORK OF OTHERS

Very seldom is research done in a vacuum. In the vast majority of cases research is built on prior work that has been documented by others, often in journal publications. Even truly interdisciplinary research done at boundary areas not previously explored, often borrow the techniques and approaches from one discipline or the other and apply them in a new field or to ask new questions. As a member of the research community it is essential that you not only know what research has been done previously but also cite that prior work as the foundation of your own when it is appropriate.

> **Student Perspective**
> "The whole of the scientific pursuit is based on openness and peer-review, and it constantly builds on previous discoveries. As the step-by-step process continues, what was learned before must be acknowledged and respected."

On the positive side, citing the work of others also helps to build your own credibility. When you have informal conversations with people in your field, write reports and publications about your research, present at a conference, or give a formal research talk on y our campus, it is critical that you discuss the background of your research area. In doing so you will need to identify who made the early findings, who established critical techniques, and who has presented research results that you have built upon or contradict your own. Depending on the specific circumstances, there are a variety of ways in which other people's ideas are credited. What is critical is that you find a way to acknowledge the work of others and distinguish it from your own. If you do not do so you are at risk of committing plagiarism.

> **Student Perspective**
> "'Word-for-word plagiarism' happens when the method of expression and sentence structure are largely maintained. 'Patchwork paraphrase' is the paraphrase that contains the language from the author without rephrase and some writer's own words. Both situations are considered plagiarism since the writer has only changed around a few words and phrases or changed the order of the original's sentences. For an acceptable paraphrasing, the information in the original sources is accurately relayed and expressed in the writer's own words."

Sometimes plagiarism is done intentionally. This is a risky proposition, especially given the techniques that are now available at universities and with publishers for identifying plagiarism. Even if you think you have found a way to cut corners by taking credit for someone else's work and get away with it, don't do it. It not only carries high risk of repercussions; it also carries a risk of luring yourself into other dishonest actions that not only jeopardize your career, but also the careers' of those around you and the integrity of the research in your field.[12]

---

[12] Ariely, D., 2012. *The (Honest) Truth About Dishonesty*, Harper Collins Publishers, New York.

Student Perspective

"This has become a bigger problem because it is so easy to access tons of information from different sources on the internet and it can be tempting to copy and paste and then rearrange and change a few words here and there in sentences. It can also be difficult to find the author information and dates of publication on some websites so some people may think they either do not need to cite the source or just simply don't do it. Another issue is the availability of online essay writing services. These services will prepare essays for students for a fee. It is plagiarism to hand in one of these essays because it is passing work off as your own that you did not do."

Sometimes plagiarism is committed because the rules are not well understood. However, ignorance of this in an academic setting will not be excused and often has severe consequences. If you are unsure how your institution defines plagiarism, look it up. Identify campus resources (e.g., a Writing Center) and look for guidelines and workshops on how to avoid plagiarism.

ASSIGNMENT 4-12:
INDIVIDUAL ASSIGNMENT – WATCHING THE CREDITS

While reading a journal article or listening to a seminar presentation, pay close attention to how prior research is credited. How is sentence structure used in combination with the citations? Is the work of others mentioned with the researcher name(s), if so is the first author indicated or the principal investigator of the research group? Does the author or speaker also cite their own work? If so, how is this done similarly or differently to citing the work of others?

Write a brief summary about your observations, including a few representative examples.

There may be some variation within your discipline, so continue to pay attention to how the work of others is credited when you are reading about and listening to research.

CHAPTER 5

Conducting Research

5.1   SCIENTIFIC HABITS OF THE MIND

Much has been written on the scientific method throughout history. Historian of science Daniel Siegel of the University of Wisconsin–Madison tells us that scientific method is a complex topic but often philosophers categorize scientific method into three idealized types: empiricist, rationalist, hypothetical.[1]

- The empiricist methodology, championed by Sir Frances Bacon during the Scientific Revolution, is based on experience, observation, and experiment. The basic idea is that generalizations can be developed from careful observations and categorizations. The basis of this method is to avoid prejudices and be guided by one's experience. An example of this methodology is taxonomy. However, this method cannot address all areas of science effectively and answer all questions we may want to pose.

- The rationalist methodology, promoted by Renee Descartes in the 17th century, is based on reasoning. The method begins with careful contemplation and consideration of how things must be. Geometry is an excellent example of this method. You begin with basic, self-evident axioms and then from there you can reason to a conclusion. If the appropriate basic principles can be found, then the rest can be developed through reason. Although its utility in isolation is limited, this is a powerful method in combination with the other methods. Prof. Siegel gives the example of the principle of conservation of energy, which is a constitutive principle of science which can be applied to a wide range of topics and can guide thinking and new experimentation.

- Hypothetical methodology is based on suppositions and conjectures. This was a method frequently used in science historically, but often denied. You imagine possibilities, try out ideas, and ask the question: If my supposition was true, what consequences would there be to things that I can observe directly? This method allows scientists to get beyond the limitations of what can be observed directly. For example, the idea of the atom came well before direct observations were possible, but the hypothesis of the existence of atoms has logical consequences that can be observed through experiments. The hypothesis must be a testable, or verifiable, in order for it to be a scientific method. Although this is an indirect method, it is very powerful.

---

[1] Siegel, D., video on "The Scientific Method," in "Professional Development," Materials Research Science and Engineering Center, University of Wisconsin-Madison, https://education.mrsec.wisc.edu/professional-development/.

If we look at the practice of science and engineering research today, we find that most often a combination of these methods is used. We can do research using a mixture of methods according to the needs of the research problem we are approaching.

ASSIGNMENT 5-1:
GROUP ACTIVITY – DIET COKE AND MENTOS

This activity explores scientific method through two sets of results.

The first is from the popular Discovery Channel T.V. show, *MythBusters*. In Episode 57: Mentos and Soda (first aired August 9, 2006), the show investigates the cause of the explosive reaction between these two ingredients. The results of the reaction can be found in numerous videos through an Internet search using the key words: diet coke and mentos. See `https://go.discovery.com/tv-shows/mythbusters/` for episodes.

The second set of results comes from the experiments in an article "Diet Coke and Mentos: What is really behind this physical reaction?" by Tonya Shea Coffey (*American Journal of Physics*, 76(6), (2008) pp. 551–557).

After viewing the *MythBusters* episode and reading the journal article, discuss the scientific method approaches taken in each. Consider the following questions.

MythBusters Episode

- What aspects of the scientific method are incorporated into the experiments presented, and what aspects are lacking?

- What hypotheses have the hosts made? Do they add new or modify their hypotheses as they continue with experimentation?

- Do you agree with the conclusions of the hosts?

- Is the question closed, or must further research be done?

Coffey Paper

- What aspects of the scientific method are incorporated into the experiments presented, and what aspects are lacking?

- What if any issues did you have with the experiments, methods, and analysis that they chose?

- Do you agree with the conclusions of the author?

- Is the question closed, or must further research be done?

Comparison of the Two Studies

- Compare and contrast the methods, experiments, and conclusions of the Coffey paper with the *MythBusters* study.

- Which of the studies better adheres to the scientific method?

- How does presentation style matter to the general public vs. the scientific community (would you choose to highlight one of these studies over the other when discussing the results with a friend or colleague)?

ASSIGNMENT 5-2:
GROUP ACTIVITY – MONTY HALL EXPERIMENT

Ken Overway at Bridgewater College Developed an activity based on an old television show called "Let's Make a Deal," hosted by Monty Hall.[2] In the show, contestants have the chance to win a big prize if they choose the correct door out of three options. After they have made their initial choice, the host eliminates one of the remaining doors that does not have a prize. At that point, the contestant is offered the choice of staying with their original choice or switching to the one remaining door. What choice should the contestant make in order to have the best chance of winning the prize?

Develop a hypothesis statement about the anticipated outcome of the relative odds of winning the prize at the end by choosing to switch vs. stay. Consider why you have decided on this hypothesis. Then collect empirical evidence, ideally by working with a partner so that you have a host and a contestant. Complete 20 trials of the game. Analyze the results and determine if you find the evidence to support or refute your hypothesis. Finally, restate or reaffirm your hypothesis based on this evidence. (Note: A thorough discussion of the background and activity are given by Overway.)

ASSIGNMENT 5-3:
INDIVIDUAL ASSIGNMENT – COMPARING THE EVIDENCE

Choose a science question that has received coverage by the media and is perhaps a subject of controversy. A number of health questions fall under this category, e.g., How does living near

---

[2]Overway, K., 2007. Empirical evidence or intuition? An activity involving the scientific method. *Journal of Chemical Education*, 84, 606. http://jchemed.chem.wisc.edu/Journal/Issues/2007/Apr/abs606.html.

high voltage power lines affect people? Does cell phone use cause brain tumors? Will ingesting large amounts of vitamin C help to prevent a cold?

Investigate your topic using a variety of sources, including web searches, online encyclopedias, and scientific journals. Can you find scientific journal articles that disagree with one another?

After completing your investigation, consider the following questions.

- What types of evidence are available on this question?

- Does the media coverage accurately describe the research reported in journal articles?

- How might inaccuracies in the media portrayal of research findings occur?

- What responsibilities do scientists and engineers have to convey their findings to the public?

- If scientific journal articles come to opposing conclusions, can you identify flaws in the methods or conclusions of these articles?

- What conclusion do you draw from your investigation, and why?

5.1.1   OTHER RESOURCES ON SCIENTIFIC METHOD

This section only touches on a very large topic that has been the subject of inquiry for centuries. For additional content, the following references are suggested.

American Association for the Advancement of Science, 1990. *Science for All Americans*, Project 2061, 183–187. See *Chapter 12: Habits of Mind*.

Niewoehner, R., Paul, R., and Elder, L., 2007. *The Thinker's Guide to Engineering Reasoning*, The Foundation for Critical Thinking, p. 57.

Paul, R. and Elder, L., 2006. *A Miniature Guide for Students and Faculty to Scientific Thinking*, The Foundation for Critical Thinking, p. 49.

Wolfs, F. "Introduction to the Scientific Method," University of Rochester, `http://teacher.pas.rochester.edu/phy_labs/AppendixE/AppendixE.html`

5.2   DEVELOPING A RESEARCH PROPOSAL

Different types of research take different time frames to accomplish. You need to be aware of this, and you should also engage your research mentor to help you develop a research project that is appropriate. Depending on whether you are embarking on a summer research experience, a master's thesis, or a Ph.D. dissertation, the scope of your research project will necessarily be quite different. Your project might be quite independent of the work of others, or other researchers may be depending on the results you produce in order to carry on their work. It may also be the case that your research is a contribution in a much larger, longer-term effort, but you should have an idea of what milestone(s) you are trying to achieve. Understanding these aspects up front are important to appreciating how your research fits into the research group you are joining and the broader field that your work connects to.

Student Perspective

"I think the most surprising thing I learned about the nature of research this semester was the time frame on which research is carried out. Although I wasn't under the assumption that research is instantaneous or that a project could be finished in a week, I never really thought about the actual time frame of it all. But, now I realize that some research projects can take years, even decades to get publishable results."

After you have the basic of the research you will be undertaking you need to be able to express it clearly and succinctly in your own words. This can be harder than it sounds because you need to have done some reading and had some conversations with your research mentor so that you can develop a good background understanding of the area of research that you are undertaking. Then you need to write out the research question in a way that everyone will understand— you, your mentor, others in the field, and people outside the field. Use the least amount of jargon possible and rework this short statement until you can hand it to someone who does not know your research area; success is when they understand what you are planning to do and can explain it back to you based on what you have written. If they can't, listen carefully to their questions because this will likely help you to identify where you may have leaps in logics or fuzziness in your explanation.

Student Perspective

"One of the former ... students advised that we should be able to write down the basic idea of our thesis on a napkin. This doesn't mean I should expect to know what results I will achieve or what questions I will answer along the way, but having a clear question in mind at the start of the project is important. Before beginning and while performing the research, it is important also to set realistic goals."

Often as part of research you will need to take a role in preparing a proposal. This may be a requirement of your degree program, part of a fellowship application, or something you work on with your research mentor to submit to a funding agency. There are a variety of styles and expectations depending on the specific proposal being written. The first step you should take is to learn about the expectations—check your program requirements, fellowship application instructions, or funding agency guidelines. The next step is to see if you can find a good example to help you understand what a succesful proposal looks like—use your network to see if you can find someone willing to share a copy of theirs. These steps will help you prepare to write a successful proposal of your own.

In some cases, you will need to frame your research with a hypothesis. If you have not been given a hypothesis to explore by you research mentor, you will have to come up with your own hypothesis after you have read the literature and immersed yourself in your research group to learn, explore, and develop your ability to come up with a new idea. Regardless of whether the idea originates from you or your research mentor, you need to make sure that someone else has not already answered this question. This means taking a deep dive into the literature (not just the last 10 years) and looking into both the literature of the specific subdiscipline in which you are engaged and other related research areas. Use more than one search engine, as each has slightly different coverage and indexing nuances. Then you can go about fashioning a testable hypothesis around your idea.

To propose this work, you will not only need a research question or hypothesis, but you will also need to have a plan for how to go about carrying out the research. Determine what you will need to do to test this hypothesis. Will you approach it with theory, modeling, experiment, or some combination? Determine the facilities, tools, materials, and background knowledge you will need. Do you have access to these or can you find a way to gain access? Develop a plan for the steps you will take to complete the research. Identify alternative strategies you will use if your initial approach does not work out as expected. All along the way, get input from your research mentor, other experts, and others in your research group.

There is a balance however, and at some point, you need to start on the research itself. This might mean conducting a preliminary experiment or making a simplified model to test out some of your ideas. If you are able to capture some data or show some initial results this can be very useful to include in the proposal. It shows that you have already been able to make some progress on your idea and that you have the fundamental skills needed to carry out the research you are proposing.

Student Perspective

"One of the tendencies I had (and still pops up at times) is that I wanted to understand all the theory, papers, and work out there first so I knew what I was doing in my experiments and not wasting time. I almost feared just running a trial or two to better understand, and often without doing so I had no chance. I received advice to spend more time in the lab. That looking for research papers was good for understanding where things have been, save time when implementing earlier found discoveries or techniques, and even inspires new ideas. However, the quickest way to see what works and doesn't will be through trial and error, learning form each experiment."

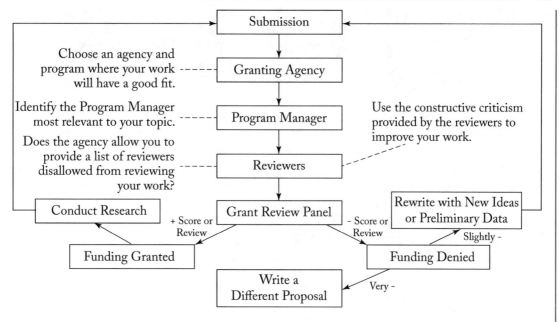

Figure 5.1: Proposal submission process.

If you are writing a fellowship proposal or working with your research mentor on a proposal to be submitted to a funding agency, it is helpful to understand how the decision process works once you have submitted the proposal. Most funding agencies and foundations follow the same general submission process and have the same basic steps. The schematic[3] in Figure 5.1 provides a general sequence for how this often occurs. Your research mentor can help you to determine if this is a good representation for the proposal process that you are currently undertaking.

5.3 GETTING STARTED AND STAYING MOTIVATED

Research is inherently challenging, but it can also be fun and exciting if you make the commitment and put in the effort. The front-end investment of time is often high and it usually requires you to do a significant amount of learning and skill building before you can make progress.

> **Student Perspective**
> "In the lab, I've been given much more independence and responsibility. This has forced me to take more initiative than I had to previously, but

---

[3] Adapted from: Barker, K., 2006. *At the Bench: A Laboratory Navigator*, Updated Edition, Cold Spring Harbor Laboratory Press, Cold Spring Harbor, NY.

also allowed me to figure things out for myself. This isn't to say that I wasn't working hard before this, but I had a little bit more of a "safety net" when doing my work. Although this extra responsibility may entail a little bit work on my part, it is also much more rewarding."

Students are used to taking classes and having an externally determined schedule of homework deadlines, project due dates, and exams. Sometimes it can be difficult to transition to research when the work schedule and deadlines are most often self-imposed. Your research mentor does not want to spend the time it would take to micromanage your research project and lay out every step for you. Initially, you will require more guidance, but quickly you need to take on this responsibility for yourself. Often students find it easiest to make progress if that have a regular work schedule for research (i.e., specific hours set aside each day when research will take place). You need to have a basic plan in mind and list of tasks that need to be accomplished, but you will also need to be flexible and adjust as the need arises and obstacles pop up.

Student Perspective

"I realized that when taking on a research project, there is no room to make excuses and if you want to successfully complete a project you must always take the initiative and go the extra mile."

Learning to work carefully and with intention is also a critical research skill. Mistakes will happen, things will break, and that will be forgiven if you own up to the issue quickly and take action to prevent it from happening again. Time and money can be wasted. The worst thing you can do is risk the safety of yourself or others, so you must be certain you understand the potential safety hazards before you take action.

Student Perspective

"The most important lesson that I've learned is that it is *always* important to be careful. I have experienced several occasions when I didn't have time to do things right, so I had to find time to do them twice. Not only does this cause lost time, but it also incurs financial costs. I must always be sure to slow down, check, double check, and then proceed. This is the most crucial skill for an experimentalist."

Even if you are handed a project concept by your research mentor at the beginning, it will not only be your responsibility to carry it out but you will also need to take it to the next level. Because the very nature of research means that it has never been done before, it often does not go quite as it is planned. Changes in the scope and direction of the research often occur as the research progresses. Sometimes things don't work out and you have to develop a new path forward. If you have been thinking about the research deeply along the way, you may have noted opportunities for other exploratory work or alternative hypotheses.

Regardless of how much time you invest and your level of perseverance, there will still be low points. Everyone struggles at some point with either getting started or staying motivated while conducting experiments, coding, writing, etc. This can happen for a variety of reasons. The beginning of a project can feel daunting because you don't know where to begin or you are afraid you will make a mistake. The middle of a project can feel like chaos sometimes, especially when the direction you had started in does not work out and you have to rethink things. And there can be struggles at the end with just getting that last bit done or writing about what you have already accomplished. These are normal experiences and they are not insurmountable. Often you can get back on track by asking for advice.

Turning the Tide

In a book I wrote for junior faculty called *Survive and Thrive: A Guide for Untenured Faculty* [Morgan & Claypool Publishers, 2010] I talk about the challenges of doing creative work like research: "I have come to believe that these ups and downs are a natural part of the cycle of any career that demands creativity. Smart creative people can't be smart and creative 100 believe that the denial we actively engage in often exacerbates the problem. So, I'll admit it to you, I have had slumps. For the most part, I too have hidden them from my colleagues. For me, I think it is primarily the fear of being accused of being a fraud—the old "imposter syndrome." But what I have discovered over the years is that, if I can find one trusted colleague that I can feel comfortable talking with honestly and assured in their ability to keep my confessions confidential, the conversation relieves much of the burden and is often enough to turn the tide and give me the means to get myself out of the slump. When I have initiated these conversations, I have also found my colleague telling me "I've been feeling the exact same way recently" or "I remember feeling the same way you are now back when." Knowing you are not alone also relieves some of the self-doubt."

Student Perspective

"In a naïve way, I previously believed that if you worked hard enough, progress will always be made. This may be true, but I have learned to gain motivation in the small accomplishments and even the disappointments that end up setting you back on track."

Often people who have not run into serious motivational slumps before are the most challenged because they have not developed strategies to turn things around. It's also the case that strategies which have previously worked for you in other situations or at other times may fail you at some point. Thus, it is helpful to have a collection of strategies that you can choose from to help you build or regain your momentum. Here are a few to consider.

- Use your calendar to block out time for specific activities. If you don't actually have time in your calendar for research and you expect to "fit it in" then you need to treat your research time more like an appointment that you must keep. The simple act of spending time on tasks will help you to make progress and get back to a productive trajectory. During these blocks of research time you should be free from distraction (devices off if possible, no additional screens or popups to distract you).

- Break larger tasks down into smaller ones, pick the task that seems easiest and do that one first, then take each one in turn.

- Create "To Do" lists where you to prioritize critical items and focus on those initially. Using the Project Management strategies are essential (discussed further in the next section), but you may have to break things down into smaller steps and provide yourself with a very specific, task oriented "To Do" list every day. Be realistic about that you put on the list given the time you have available and pause to take some pleasure in crossing off an item once it is done.

- Team up with a peer working on a similar kind of project and help each other to set goals and keep them.

- Sometimes it is as simple as just getting started. Set a timer for 25 minutes. Press start and work until the timer goes off. After the timer goes off, take a short break—ideally you will get up and stretch if you have been sitting or sit down and rest if you have been standing, you might also choose to listen to a piece of music or grab a refreshment— but this is also a timed event. Your break should be a short one (3–5 minutes) and then reset the timer for another 25 minutes of work. This is a time management strategy called the Pomodoro Technique and there are apps available, although a simple timer is all that is really needed.

The most important thing to remember is that what gets you started and what keeps you motivated is a personal thing. There are a number of different factors that may influence your motivation, including the variety in the tasks you undertake, the flexibility you have in the work, and the level of responsibility you are given. Consider what factors influence your motivation and you may be able to talk to your research mentor about making modification so that you can maximize these aspects.

Ultimately, you are responsible for understanding what works best for you to keep motivated and then use that strategy. If it stops working, try out a different strategy. The ones given above are just a small sample, if this is a topic you want to learn more about, there are plenty of resources available. Consider looking into campus workshops that might be available and books that might be useful to you. I found that *The Power of Habit* by Charles Duhigg and *Drive: The Surprising Truth About What Motivates Us* by Daniel Pink to have some interesting and useful strategies.

5.4   PROJECT MANAGEMENT

Project management generally involves ideas, people, resources, and time. Often in engineering research you are handed an idea as you walk into a project, however, this idea is likely to be incomplete or even contain fundamental flaws that you will only be able to discover as you engage with the research over time. As a result, you will need to manage the ideas as you go—altering assumptions, reframing hypotheses, developing ideas to get yourself around roadblocks. These modified and new ideas may be ones you generate yourself, or they may be ones that you generate with your research mentor, your collaborators, and/or your research group. For your research project you will need to track and manage these ideas through the evolution of the project and check in with your research mentor about how the ideas are modifying your project over time.

> Ideas—You will have to manage ideas as they shift and change over time when new information and data become available. Write these ideas down. Forcing yourself to articulate them will help to refine your thinking. Revisit this over time and think about how things have changed with new information you now have. You may in fact have to shift your project or redefine your hypothesis.

The ideas themselves have to be actualized by someone—usually you—but often in engineering research there is collaboration and teamwork involved. The people involved may be simply you and your research mentor or it may include you, your research mentor, and many other people involved with a large collaborative project. Usually it is somewhere in between—involving you, your research mentor, other students in your research group, and maybe a key collaborator. In some cases it may be as simple as working with another graduate student or technician in the group to learn how to perform a task or us a piece of equipment that you will need. If everyone works with the same research mentor it is often fairly straightforward; in other cases you may need to discuss with your research mentor about how to navigate getting the help you need. A more involved people management aspect of research is working in a more supervisory role, for instance with an undergraduate research assistant.

> People—The people management aspects of a project involve both managing yourself and the interactions with others you are working with, including your interactions with your research mentor. Take responsibility and determine how best to interact in order to move your research forward. Strive to develop productive working relationships.

You may think it is your research mentor's responsibility to manage you, but it is also your responsibility to manage your relationship with your research mentor. As discussed previously in concepts surrounding "mentoring up," you need to take ownership in managing this relationship so it is the most productive possible. Also, you need to understand that your research mentor has a broad range of responsibilities and you need to take into account the time they will have available to devote to you. Recognize that the amount of time available and when it is available may not be exactly when it is most ideal for you, so you will need to plan ahead and make adjustments. This means considering their availability for meetings, how quickly they can respond to questions that arise, and how long it will take them to review drafts of your writing.

As you progress in your graduate degree program you will have a committee that is made up of several faculty who you will interact with. These individuals may be variously involved with details of your research, evaluation of your progress, your preliminary examination (the written and/or oral exam that determines your readiness to become a dissertator in the Ph.D. program), and the final defense of your thesis/dissertation (i.e., the oral presentation you make about your research at the end of your degree program). You should be able to determine the specific requirements of your degree program and who is involved at each stage from handbook information provided to you by the program and conversations with your research mentor. When working with your committee and scheduling time for key events such as the final defense of your thesis/dissertation, you must also take into account their schedules and availability. You will need to know when they will be in town or otherwise available (via phone call, video conferencing, or some other electronic means) and how that relates to the timeframe in which you would like to finish as well as the deadlines set by your university. It will also be important for you to ensure that you planning timeline provides enough time for them to give you feedback and read whatever written materials you are expected to provide them. This requires advance planning for the semester you expect to complete a major milestone and the semester you plan to finish your degree.

In order to get your project done you will likely need some resources. Even a theoretician, who only uses pencil and paper to do their research, needs resources (e.g., a paycheck). As a principal investigator some day you would have to worry about where the money comes from (i.e., getting grants and contracts) and how to manage the expenditures of money responsibly to get the research project done. As a student you will need to worry about money in a couple of ways too. You need money to live off of, possibly this comes through the research grant or a fellowship so it is less to worry over, but you may have to do other non-research work like a teaching assistantship that will help to pay your bills and cover your tuition. Needing to earn income through a non-research related position will inevitably cut into the time you spend on research. At some stage you may decide to seek other alternatives like a student loan so there are fewer time constraints.

Resources—From personal finance to the resources required of your research project, you will need to understand the resources available to you. Determine where the funding for your project comes from and what constraints exist regarding that funding (how much is available, what it can be spent on, when it expires, and what proposals are being planned for future research funding related to your project).

There are variety of different kinds of resources that you may need. It may be access to computational or equipment time; it may be samples, supplies and consumables. It may be bench space to conduct the work or lab space to build a structure for your research. Determining what these resources are and their availability will involve a number of questions that your research mentor can help you to answer. These pieces of information will impact your schedule, planning, and time to degree.

> **Student Perspective**
> "I will be writing a substantial amount of code, and I haven't quite decided if I should write it on my personal laptop in an integrated development environment or physically move myself to my office .... While working on my own computer seems "easier," it's probably a more efficient use of time to move myself to a dedicated *working* environment. With a schedule and place in mind, I hope to condition myself to work well and make a consistent amount of progress on a weekly level."

The aspect of your project that is often in most short supply is time. Your project does not happen in isolation and it is not the only thing vying for your time. Early in your career you are juggling time spent on coursework with time spent on research, not to mention personal time. All are important. Certainly, the coursework and research will both have deadlines and expectations for progress associated with them. At some point in your graduate studies you will likely be done with coursework and "only" have to do your research, but at this point you are likely to have other responsibilities you are juggling too. This may involve supervising and training others in the research group for instance. You will also need to balance the time of doing the research with presenting and writing about the research for conferences, journal publications, or your thesis/dissertation. As you approach the end of your degree you will also need to devote time to finding the next position—as an undergraduate you may be applying to graduate school, as a graduate student you will be applying to postdoctoral or permanent positions in academia, government labs, or industry. The application and job search processes takes time and needs attention while you are busy finishing up the degree you are currently working on.

Time—Time for coursework, research, and yourself. You need to be a healthy person to be an effective researcher. Get enough sleep, exercise, and have time for your family, friends, and a hobby. Being your best at research will be easier if you are a healthy, happy person. You will be more easily able to achieve this if you employ some time management techniques. You also need to communicate about time with your research mentor: your work schedule, your ability to achieve deadlines, your need for modification when other obligations and personal needs arise, your plans for vacation, etc.

Time on task is a critical component to seeing any project through to completion. But don't be fooled, just spending lots of time on something does not mean that you are making progress. Not only do you need to monitor the time you are spending, but also how you are spending that time, and if this time if productive. In order to do so effectively, you need a good project plan to guide the use of your time and measure your progress against.

> **Student Perspective**
> "In my opinion, to be a successful researcher you don't simply need to produce good research (obviously a huge plus), but you need to do it efficiently. Splitting projects up into manageable chunks not only made it seem less daunting, but also saved me a significant amount of time in the long run. Putting in the effort to lay out a solid plan before doing the work will save a lot of time, allowing me to focus on the difficult tasks. Writing down my thoughts in a notebook will save time by not having to produce the same thought twice, or completely wasting time by forgetting the thought altogether. Similarly, learning to be more *consistent* about writings will save time in the same way. Finally, as I continue to grow as a researcher, I'll need to learn when to cut my loses and switch the direction of a project. Every researcher, good or bad, will reach a point where they need to change a project; a good one will be *efficient* and not waste time trying to save the idea."

In a section titled "Professional Success: Project Management" author Prof. Mark Horenstein of Boston University gives three "Laws of Time Estimation"[4]:

1. "Everything takes longer than expected."

2. "If you've worked on something similar before, estimate the amount of time required to finish the task. The actual amount of time required will be about four times longer."

3. "If you've *never* worked on something similar before, estimate the amount of time required to finish the task. The actual amount of time required will be equal to the next highest time unit. (For example, something estimated to take an hour will take a day; something estimated to take a day will take a week, etc.)"

Although you might think that these "Laws" are an exaggeration, they can actually fit reality better than you expect. Certainly, in research it is often true that it takes longer to accomplish what you set out to than you would have anticipated. When by definition you are

---

[4]Horenstein, M. N., 1999. *Design Concepts for Engineers*, Prentice Hall, Upper Saddle River, NJ, p. 86.

doing something that has never been done before, it can be very challenging to judge how long it will take! Even for the mundane everyday tasks that you will have to handle, you probably already appreciate from experience that you have to give yourself some cushion. Back in the day when I had a paper calendar where all of my appointments, deadlines, and tasks were written, the first page contained the following quote from an anonymous source: "WARNING—Dates in calendar are closer than they appear!!" I found it often to be true, particularly when working on an immovable deadline for a research proposal.

However, this does not mean that we should throw all planning out the window. In fact, planning is what makes it all more manageable.

5.4.1   PROJECT MANAGEMENT TOOLS

Often what is critical in project management is not simply understanding all the components of a project, but how they relate to and are dependent on each other. In a research setting this is important whether the work be experimental, analytical, theoretical or some combination. It may also be the case that your project is dependent on other projects or a part of a larger effort involving a number of other researchers. In that case you would not be responsible for the overall project management, but you would still have to manage the aspect you are responsible for and meet the necessary deadlines so that the rest of the project can go as planned.

The project must also not be over-constrained.[5] In other words, the scope must be within your capabilities (or, more likely, the capabilities you will be developing); the resource requirements must be within the budget, equipment, and facilities available; and the schedule must be reasonable, given the amount of time you are able to invest in your research.

In order to fully understand your project and develop a project plan there are a some basic steps that you can take.[6]

1. First, clearly identify the project with a succinct problem statement.

2. Identify the tasks that will need to be accomplished in the course of the project with as much specificity as possible. Define specific milestones for the project that will allow you to measure your progress.

3. State the objective of each task so that its purpose is clearly delineated. You can think of these as deliverables.

4. Identify the people who will be involved with each task. You may perform some of them independently, but other tasks may require the participation of others (e.g., training), or rely on someone else to provide you something in order to complete the task.

---

[5] Kendrick, T., 2009. *Identifying and Managing Project Risk: Essential Tool for Failure-Proofing Your Project*, 2nd ed., American Management Association, New York.
[6] Adapted from Ullman, D. G., 2003. *The Mechanical Design Process*, 3rd ed., McGraw Hill, Boston, MA.

5. Estimate the time it will take to complete each task and identify the distribution of the time across different phases of the project.

6. Estimate the funding, equipment, computing time, or other resources that will be required to complete each task.

7. Identify the sequence of the task, noting interdependencies between tasks (precessors and successors). Note which tasks can be done in parallel. Identify any externally imposed deadlines.

A number of tools are at your disposal. The most basic is a simple timeline. You have a target end date to work toward and various milestones to meet along the way. An example is given in Figure 5.2. However, a simple timeline does not provide information about how one might be working on multiple objectives simultaneously or show how long each task will take. A Gantt Chart, like the one shown in Figure 5.3, provides additional information about when project activities are occurring and shows how they overlap. Additional modification to the Gantt Chart can also indicate dependencies and relationships between items, e.g., data must be collected before analysis can be undertaken, but to fully explore the interdependencies, you may want to create a structure matrix like the one shown in Figure 5.4. Each task is assigned a letter name and each row of the chart identifies the other tasks on which each one is dependent.

At some point, however, you must stop planning and start doing. The fact is, no matter how much time and effort you put into your planning, some changes will inevitably arise. Your ability to handle these changes and quickly develop a revised plan will depend on the thoroughness of your original planning effort, built in cushion within the plans, and flexibility with at least one of the constraints of scope, schedule and resources. As Kendrick summarizes, "Your primary goal in managing project constraints is to remove, or at least minimize, the differences between the project objective and your project plan, in terms of scope, schedule, and resources.[7]"

ASSIGNMENT 5-4:
INDIVIDUAL ASSIGNMENT – COURSEWORK TIMELINE

Create a timeline for completion of the coursework you have remaining in your degree program.

---

[7]Kendrick, T., 2009. *Identifying and Managing Project Risk: Essential Tool for Failure-Proofing Your Project*, 2nd ed., American Management Association, New York, p. 128.

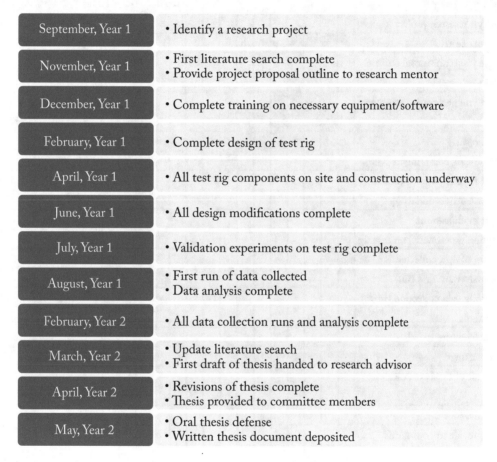

September, Year 1	• Identify a research project
November, Year 1	• First literature search complete • Provide project proposal outline to research mentor
December, Year 1	• Complete training on necessary equipment/software
February, Year 1	• Complete design of test rig
April, Year 1	• All test rig components on site and construction underway
June, Year 1	• All design modifications complete
July, Year 1	• Validation experiments on test rig complete
August, Year 1	• First run of data collected • Data analysis complete
February, Year 2	• All data collection runs and analysis complete
March, Year 2	• Update literature search • First draft of thesis handed to research advisor
April, Year 2	• Revisions of thesis complete • Thesis provided to committee members
May, Year 2	• Oral thesis defense • Written thesis document deposited

Figure 5.2: A simple timeline identifying critical milestones in a fictitious Master's degree research project.

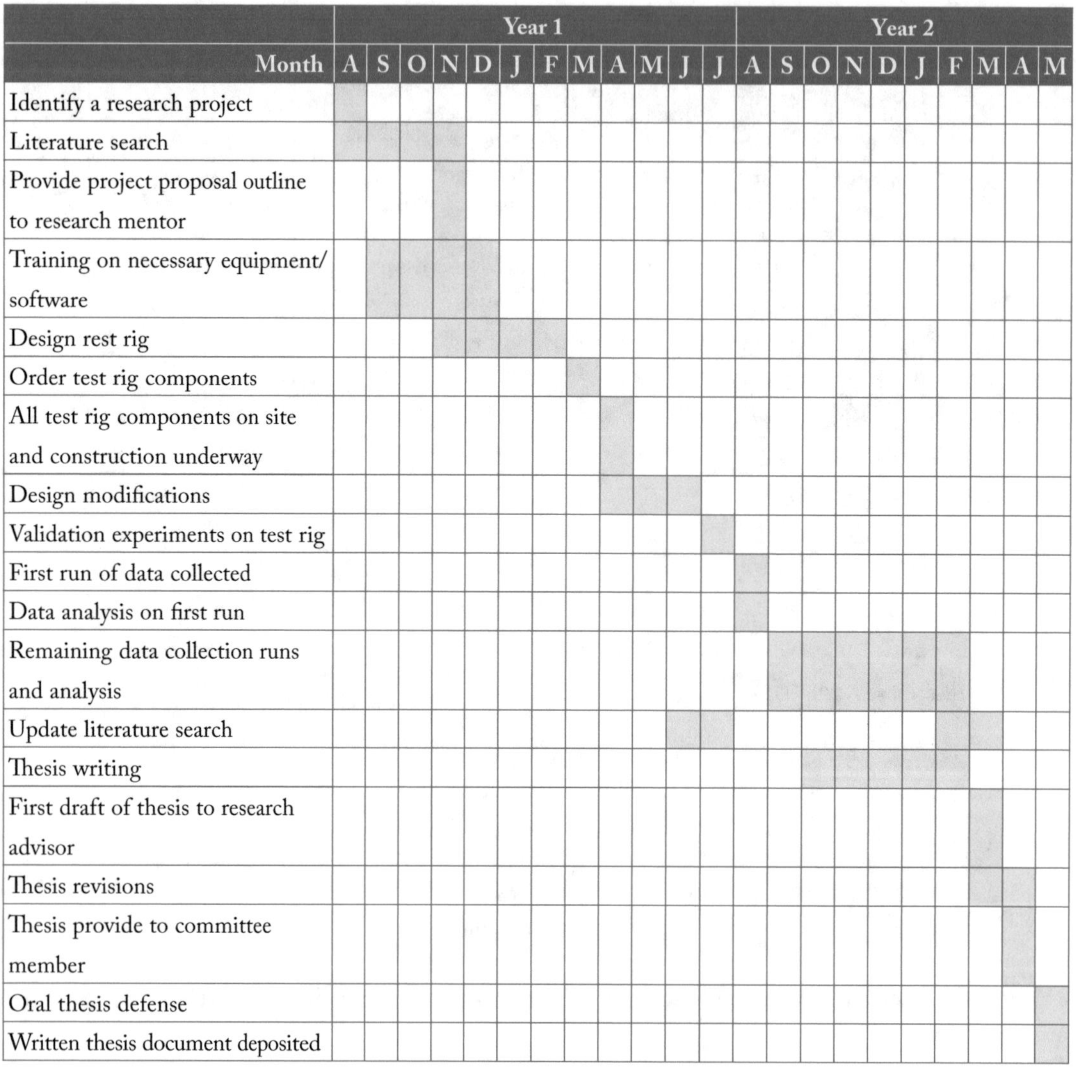

	Year 1												Year 2									
Month	A	S	O	N	D	J	F	M	A	M	J	J	A	S	O	N	D	J	F	M	A	M
Identify a research project	■																					
Literature search	■	■	■	■																		
Provide project proposal outline to research mentor					■																	
Training on necessary equipment/ software		■	■	■																		
Design rest rig				■	■																	
Order test rig components								■														
All test rig components on site and construction underway									■	■												
Design modifications										■												
Validation experiments on test rig												■										
First run of data collected													■									
Data analysis on first run													■									
Remaining data collection runs and analysis														■	■	■	■					
Update literature search										■												
Thesis writing																	■	■	■			
First draft of thesis to research advisor																			■			
Thesis revisions																				■		
Thesis provide to committee member																					■	
Oral thesis defense																						■
Written thesis document deposited																						■

Figure 5.3: A Gantt Chart showing when activities in the time line are occurring.

TASK		A	B	C	D	E	F	G
Design test rig	A	A						
Order test rig components	B	X	B		X			
Test rig construction	C	X	X	C				
Design modifications	D			X	D	X		
Validation experiments on test rig	E			X	X	E		
First run of data collected	F					X	F	
Data analysis on first run	G						X	G

Figure 5.4: **A structure matrix showing the interdependencies of project tasks.**

ASSIGNMENT 5-5:
INDIVIDUAL ASSIGNMENT – RESEARCH PROJECT GANTT CHART

Develop a Gantt Chart that will help you to plan and complete your research project by the deadline you have chosen. Consider all competing demands on your time, such as your course-work requirements. Consult your research mentor with a draft version and seek input about whether or not your planning is reasonable.

5.5   SCHEDULING COMMITTEE MEETINGS

It's likely that you have a busy schedule on a day-to-day basis given your courses, research, and other personal obligations. The faculty members you interact with also have busy schedules and are usually quite busy at just the time of year that you may need their time and attention most. With your research mentor it is ideal to have regular times when you have the opportunity to interact so that you do not need to schedule each individual meeting that you two will have.

Your interactions with other faculty will require individual scheduling of meetings and in the case of a committee meeting, it will require juggling the availability of multiple very busy people. In these cases, it is critical to plan ahead. If you know you will need to give an oral presentation to your committee in order to complete your degree requirements in the last two weeks of the semester, then you should start the planning process more than a month in advance so that you are certain you can find a time when everyone can meet. There are a variety of strategies to go about this planning process, but I think the smoothest interactions take place using the following steps.

- Consult any degree deadlines that apply and ensure that you know when the latest acceptable meeting date will be.

- With the guidance of your research mentor, determine what weeks would be appropriate for the meeting/presentation and identify the length of time that you will need to schedule (this could be anywhere from 1–3 hour time block depending on the specific circumstances).

- Contact the individuals on your committee independently (by email or a personal visit to their office) to determine which dates they will be in town and generally available during the target time frame that you are interested in.

- Compare their availability to your own and make a list of all the potential time blocks that fit all the criteria.

- Using a scheduling tool such as Doodle or WhenIsGood can be helpful in narrowing down workable options. Alternatively, you can list the potential day/date/times in an email and request they respond with all options that will work for their schedule.

- The above steps should be undertaken as quickly as possible because as time passes more and more obligations will fill up your research mentor's and committee members' calendars. If someone does not respond to an email request, go to their office to ask about their availability.

- If your scheduling attempt does not work the first time, you will have to start over again and identify different dates with your research mentor. If that is not an option, it may be possible to have one member join the meeting by phone or video conference call if they are out of town, or have you meet with them independent from the remainder of the committee. If none of these options work you may have to determine if a committee member can be substituted with a different person. For all these reasons, it is a very good idea to start the scheduling process early.

5.6   NAVIGATING ROADBLOCKS AND OBSTACLES

Inevitably there will be roadblocks and setbacks. I can't think of a single research project that did not have at least small issues arise. It is the fundamental nature of research, particularly as you investigate areas that are at the cutting edge of a field. You may find it helpful to not only keep your overall research goal in mind, but also to break this large goal down into smaller sub-goals that you can more easily focus on when you run into bumps and hurdles.

Research can create more unknowns in the initial planning process and therefore more points at which the plan must be revised. In some cases, the constraints are quite hard. For example, when working on a research contract, certain deliverables are expected within a specific time frame. The time frame might be modifiable, but usually the deliverables are quite fixed in nature. In other more open-ended research projects, there may be more room for modification of the research objectives, particularly if preliminary findings uncover what has the potential to be a more fruitful line of inquiry. Even if you are not able to pursue these new ideas right away, it is helpful to write them down so you can explore some of them later or use one as a basis for a new research proposal. It is important to always be open to opportunities, especially when faced with challenges. "Opportunity management also may result in a more interesting, more motivating project....[8]"

> **Student Perspective**
> "My proposal had very lofty goals, and I knew that from the outset. I was upset on my progress at first, but now I realize that for every problem I run into, I'm learning more and more about the subject matter. The whole "if at first you don't succeed…" motto has some clear consequences …. I know now that the full scope of my proposal will not be represented in my final product."

In some cases, you will have to re-propose or re-negotiate your project with your research mentor, but you can do so in a way that sets you up for success. To do so, you must develop a revised plan with an appropriate scope, resources, and schedule which will allow you to accomplish the research. Before deciding to simply reduce the scope, think creatively about how the scope might be shifted to take advantage of what you now know about the project. If reducing the scope is still required, determine the essential outcomes of the research before making any cuts. When considering resources, think creatively about what other resources might be available to you. Examples range from applying for a small seed grant to finding an assistant to take some of the work burden. Schedule modification will require a careful analysis of critical path activities and interdependencies in the project. Additional aspects of the project may need to be conducted in parallel and/or some time frames tightened up to allow for more flexibility in a different part of the schedule.

Research is inherently challenging because you are trying to do something that has not been done before. You will inevitably run into roadblocks and obstacles and have to think of creative ways to get around, over, or through them. This is fundamentally a part of the process and sometimes these obstacles can be the very thing that lead you to an unexpected and fruitful outcome.

> **Student Perspective**
> "The most surprising thing that I learned about research this year was that research could be easily delayed by sudden problems in the laboratory. One of the largest aspects that research deals with is getting the research systems to not only work, but to work continuously over a certain period of

---

[8]Kendrick, T., 2009. *Identifying and Managing Project Risk: Essential Tool for Failure-Proofing Your Project*, 2nd ed., American Management Association, New York, p. 133.

time. Working in a research lab on campus has shown me the multitude of failures in machinery, computer code, and other systems that can slow down the efficiency of a lab and delay the research mission. A failure in a critical system of the experiment could stop work in the entire laboratory until the problem with that one system is addressed. I believe that a lot of people have misconceptions about the research process, and they believe that scientists and engineers just turn buttons inside of a control room and research happens. Being exposed to research on campus, has shown me the "dirty" side of research, where hours upon hours are spent fixing machine failures, designing new systems to replace faulty processes, and brainstorming how to fix a problem that you encounter that you originally thought not possible.... While, these problems can prove detrimental to the research mission, I believe that experiencing and overcoming these problems is one of the general responsibilities of being an engineer in research. Encountering problems allow you to design new more efficient systems; as well as take a step back and think about your research in a different way."

Consider the following challenge as an example that came up in my own research group a few years ago. The supplier we had used previously to make a photolithographic master changed its focus and was no longer supplying what we needed. Having a new master was a key component that we required in order to test out a new design critical to the successful completion of our project. We had to look for options and develop a plan to find a way to have a new master made and do the experiments we had planned. In our lab we use the photolithographic master to make polydimethylsiloxane (PDMS) stamps that can transfer a protein pattern onto a substrate. This allows us to seed and culture cells in specific pattern designs. The basics of the process are illustrated in Figure 5.5.

How could we get a new photolithographic master made? First, we considered both off-campus and on-campus sources. We looked for a new commercial supplier—a Google search can be helpful if you know the right words to search on, but it can also be helpful to look at recently published journal articles using the same technique to determine who their supplier was. We quickly found two potential commercial options and began to make inquiries about whether they could meet our specifications and how much the cost would be. While doing this, we also wondered if there was another lab on campus using the same method. We turned to the professional network of our research group to see if one of us knew someone who could help. It turned out that one of my students knew a student in another lab who was doing something similar—their research mentor was a colleague of mine, so it was easy to ask for advice. They made something similar in their lab and offered to try to make what we needed. Additionally, we thought about whether this was something we could easily make ourselves. There are methods

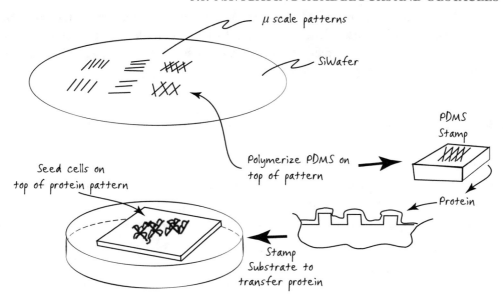

Figure 5.5: Use of a photolithographic master to create patterned stamps for protein transfer onto a substrate and subsequent cell growth on the protein.

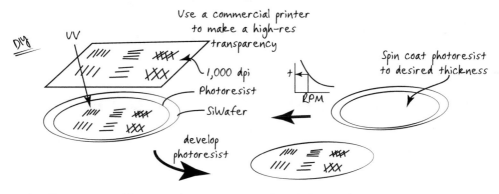

Figure 5.6: Do-it-yourself approach to creating a photolithographic master.

published in the literature and we could purchase the photolithographic master solutions. A sketch of our do-it-yourself approach is shown in Figure 5.6.

We quickly ended up with several options: two potential commercial suppliers, a colleague willing to try to help us out, and a method for how to approach it if we decided to make it ourselves. What had seemed like a big obstacle—the loss of a supplier for a critical research component—had turned into a solvable problem! In the end we tried one from our colleague's lab, but it did not quite work with our subsequent processing steps, so we used one of the other commercial suppliers and the roadblock was cleared. There are several lessons to take away from this example: there is often more than one solution to a problem; there is often more help with an issue than you might have initially suspected; and pursuing multiple paths simultaneously may produce alternatives for you to choose from.

ASSIGNMENT 5-6:
GROUP DISCUSSION – FINDING THE RESOURCES YOU NEED

You have discovered that the key to your research is getting access to a "Widget Measurement Device." It is critical to the successful completion of your project but your research group does not have one. Develop a plan to find such a device and get the measurements you need. Consider the following questions.

- How would you go about finding such a device?

- How can you get access?

- What if it does not exist on campus?

- If you have to borrow someone else's widget, what can you give in return?

ASSIGNMENT 5-7:
INDIVIDUAL ASSIGNMENT – OPPORTUNITY MANAGEMENT

Reconsider your research plan in light of an actual or imagined roadblock and look for alternate opportunities.

5.7   RESEARCH ETHICS (ERROR, NEGLIGENCE, MISCONDUCT)

Research is conducted by human beings, so human failings inevitably enter into the picture. Sometimes the failing at play is simply error: people make mistakes. The defining moment is when you discover the error and decide what to do about it. Unfortunately, sometimes an otherwise honest person might be inclined to hide the error, which is the big mistake. The temptation to hide it may come from feelings of shame or concern over the repercussions but hiding an error crosses the line between making a mistake and being dishonest. In my own research group, I stress with students that it is critical for them to let me know when they make a mistake or discover an error. The sooner we know about it, the sooner we can resolve things. For instance, if a piece of equipment gets broken accidentally, we want to know right away so we can repair it before it is needed again, and we'd like to determine what went wrong so we can prevent it from happening again. If the error occurred in how an experiment was run for instance, we want to know as soon as that comes to light so that we can determine the best course of action moving forward, e.g., repeating the experiment using the correct protocol and revising our training procedures so that this type of error does not happen in the future.

If the error ends up not getting caught until after publication of the results, there is still an opportunity to fix things. Journals publish errata to correct errors discovered after publication and corrigenda to correct errors made in the publication process (like a typographical error in an equation). In the most extreme cases, when the results and conclusions are significantly altered by the issue, the article may be withdrawn from publication if the editor(s) agree that this is the best course of action. You may think this is a horrid outcome, but your scientific colleagues would rather know that the error exists than to base their ongoing work on something that is flawed. It will waste your time and waste the funding supporting your research if you pursue a path that someone has discovered is wrong but has not made the effort to report it. When Prof. Pamela Ronald discovered that two papers had incorrect data and conclusions because of lab errors associated with the mislabeling of a bacterial strain and an unreliable protein assay, she decided to announce it publicly at a Keystone Symposia Meeting and retract the papers.[9] Her colleagues applauded her forthrightness about the issue. Honest mistakes happen, and they should be acknowledged when they are discovered. "Scientists who make such acknowledgments promptly and openly are rarely condemned by colleagues.[10]"

In the hierarchy of bad things happening, negligence falls somewhere between error and misconduct. Haste, carelessness, and inattention must be treated more harshly than an error. In the case of negligence someone is cutting corners. This is not good science, and the outcomes may have very negative ramifications for engineering research broadly. Money, structures, and people's lives may be put at risk. On the professional side, not only is the negligent individual placing their own reputation at risk but they can also damage the reputations of their colleagues and researchers as a whole. Further, "By introducing preventable errors into science, sloppy or negligent research can do great damage—even if it is eventually uncovered and corrected.[11]" Sometimes we don't realize that there was a problem until we try to repeat the work later. In an anonymous survey of National Institutes of Health researchers, 6% of the respondents admitted to "Failing to present data that contradict one's own previous research.[12]" This is really problematic. Even though you and others are reading the literature critically, everyone is relying on that prior research so that they can build upon it.

---

[9]Grens, K., 2015. Self correction: What to do when you realize your publication is fatally flawed. *The Scientist*, December, 29(12).

[10]*On Being a Scientist: Responsible Conduct in Research*. National Academies Press, 1995.

[11]*On Being a Scientist: Responsible Conduct in Research*, National Academies Press, 1995.

[12]Martinson, B. C., Anderson, M. S., and De Vries, R. 2005. Scientists behaving badly. *Nature*, 435.7043, 737.

Misconduct is when the action crosses into the realm of intentionally deceptive. "Making up data (fabrication), changing or misreporting data or results (falsification), and using the ideas or words of another person without giving appropriate credit (plagiarism)—all strike at the heart of values on which science is based.[13]" A meta-analysis conducted by Daniele Fanelli "...found that, on average, about 2% of scientists admitted to have fabricated, falsified or modified data or results at least once—a serious form of misconduct by any standard [10,36,37]—and up to one third admitted a variety of other questionable research practices including 'dropping datapoints based on a gut feeling,' and 'changing the design, methodology or results of a study in response to pressures from a funding source.'[14]"

So why do people do it? The heart of the matter is that research can involve "...intense competition, and is further burdened by difficult, sometimes unreasonable, regulatory, social, and managerial demands.[15]" But the consequences of succumbing to the temptation can be severe. In many of the public cases reports, careers have been ruined. In some cases people have even gone to prison.[16]

For students, I find that their choice to undertake an act of misconduct is frequently an issue of time. Things don't go as planned, it takes longer than anticipated, deadlines are approaching, graduation is near, a job is waiting....... then someone gives into the temptation to cut corners by fabricating, falsifying, or plagiarizing. Maybe they did it before and got away with it the first time, or even the second time, but at some point these indiscretions will get caught. What then? It can be career ending. Years of effort invested are now wasted. As an example, discovering plagiarism in one Mechanical Engineering master's thesis, led to Ohio University to conduct a broader investigation and ultimately "...taking action against 39 mechanical engineering graduates.... It has ordered them to address plagiarism allegations involving theses dating back 20 years or risk having their degrees revoked.[17]"

In some cases, the consequences can be absolutely tragic. The work and research that engineers do is intrinsically connected to society and often directly related to safety, healthy, and environmental issues.

---

[13]On Being a Scientist: Responsible Conduct in Research, National Academies Press, 1995.

[14]Fanelli, D., 2009. How many scientists fabricate and falsify research? A systematic review and meta-analysis of survey data. *PloS One*, 4(5):e5738.

[15]Martinson, B. C., Anderson, M. S., and De Vries, R., 2005. Scientists behaving badly. *Nature*, 435.7043, 737.

[16]Rogers, G., 2014. Former ISU scientist who faked AIDS research indicted, *Des Moines Register*. https://www.desmoinesregister.com/story/news/crime-and-courts/2014/06/19/former-isu-scientist-who-faked-aids-research-indicted/10881781/.

[17]Tomsho, R., 2006. Student Plagiarism Stirs Controversy At Ohio University, *Wall Street Journal*. https://www.wsj.com/articles/SB115560632839035809.

Most engineering professional organizations have a code of conduct for their members. The National Society of Professional Engineers (NSPE) Code of Ethics for Engineers includes the following **Fundamental Canons**.[18]

"Engineers, in the fulfillment of their professional duties, shall:

- Hold paramount the safety, health, and welfare of the public.

- Perform services only in areas of their competence.

- Issue public statements only in an objective and truthful manner.

- Act for each employer or client as faithful agents or trustees.

- Avoid deceptive acts.

- Conduct themselves honorably, responsibly, ethically, and lawfully so as to enhance the honor, reputation, and usefulness of the profession."

These seem sensible. Engineers should make the world around them a better and safer place for the benefit of society. If you keep this underlying motivation in mind as you conduct your engineering research, you will be much more able to do it in an ethical manner.

5.7.1 MISCONDUCT CASE STUDIES AND THE D.I.S.O.R.D.E.R. FRAMEWORK

Lisa Newton, Professor of Philosophy and author of *Ethics in America* proposes a decision procedure for dealing with ethical dilemmas that I find very useful in applying to both case studies and real-life situations. The basic approached is summarized.[19]

"...as participants and decision makers, we should organize our **options** in the situation—what alternatives are really open to us? and note the probable **outcomes** of each. What, in this situation, is it possible, and reasonable, for us to do? And what will be the likely results of each of those choices? Which of the outcomes on the list are totally unacceptable? They should be eliminated, and the rest left for further consideration at a later stage."

The acronym developed by Newton is D.I.S.O.R.D.E.R. where we: **Define** the dilemma we are trying to address, **investigate or identify** the necessary information, identify all of the **stakeholders** that are impacted by the dilemma, explore the **options** available to us, spend time considering the **rights and rules** associated with the issue and the individuals involved, make a

---

[18]National Society of Professional Engineers, "Code of Ethics," http://www.nspe.org/resources/ethics/code-ethics.

[19]Newton, L. H., 2002. Doing Good and Avoiding Evil, Part I. Principles and Reasoning. http://www.rit.edu/\simw-ethics/resources/manuals/dgae1p6.html.

decision or determination about what actions should be taken, then **evaluate** the effects that will likely be a consequence of that decision, and at the end take time to **review and reconsider** to make sure we have made a decision that is reasonable. The D.I.S.O.R.D.E.R. frameworks provides a logical and yet flexible decision making procedure that can be used in a variety of situations.

In applying the disorder framework we focus on the options available, the rights and responsibilities of the parties involved, making a preliminary decision, considering what the effects of that decision will be, and then reconsidering. It is intentionally an iterative process that asks you to do the thought experiment of what will happen as a consequence of the decisions that you take, and then to reconsider the situation with those ideas in mind. Once you have been through the O.R.D.E.R. portion of the framework at least twice you have probably gotten to the point where you have settled on a reasonable course of action where the probable results are in alignment with the ethical quandary being and dealt with.

Let's practice by considering a simple case example that came up a number of years ago in a class I taught. The students were assigned to write a review paper. The first draft went through a peer review process with the other student colleagues in the class serving as the reviewers. When reading the paper he was assigned to review, one of the students went to Wikipedia to read some more background on the topic because it was unfamiliar to him. However, in the process he discovered that two paragraphs of the paper's introduction were directly copied from Wikipedia by the author of the paper. Quotations were not used and Wikipedia was not cited.

Let's apply the D.I.S.O.R.D.E.R. framework to this case. First, what is the student's dilemma? He suspects plagiarism and must decide what to do. He's effectively already done part of the investigation by determining that there was copying from Wikipedia, but he might also want to check the syllabus and/or the University definition of plagiarism to see if this situation fits. In this case it certainly fits the definition of plagiarism without question—the paper's author has tried to pass off the work of someone else as their own. It does not matter if it is freely available on the internet, the author did not use quotes or paraphrase the source by putting it in their own words, and author did not cite the source. Who are the stakeholders? Clearly the author of the draft that is being reviewed is a key stakeholder, but so is the student who has identified the suspected plagiarism. The instructor for the course is an obvious stakeholder, but also the other students taking the course are stakeholders too because they are all expected to do the same assignment. More broadly, you would consider the University as one of the stakeholders as well.

The student who discovered the plagiarism has a few options available—say nothing, say something to the author, or say something to the instructor. If he looks into the University of Wisconsin-Madison policy that spells out the rules associated with academic integrity, he will see that the University states that "Students are expected to uphold the core values of academic integrity which include honesty, trust, fairness, respect, and responsibility.[20]" With this in mind, the "say nothing" option could potentially jeopardize him, so that does not seem like a good option. Alternatively, he could decide to talk with the author and clarify the University policies on plagiarism, but this is not really his job and it could result in a backlash from the other student that may be challenging to handle. In fact, the University website on "Academic Integrity" goes on to say that "As a member of the UW-Madison community, it is your responsibility to help uphold the integrity of the university. If you suspect a classmate is cheating or committing another type of academic dishonesty, notify your instructor, professor, or teaching assistant. Remember that it is not your responsibility to investigate this. It is the job of the instructor to determine if misconduct occurred. All you need to do is report what you heard or saw." After reviewing and reconsidering the options, the student who conducted the review decided that the best course of action was to contact the instructor who then went on to take actions in accordance with University policy.

ASSIGNMENT 5-8:
GROUP ACTIVITY – AN ETHICS CASE STUDY ON DATA FABRICATION

Consider another case study that will allow you to further apply the D.I.S.O.R.D.E.R. framework. Many years ago a colleague contacted me about one of their students asking for my thoughts on what actions they could and should take in a suspected research misconduct case. My colleague had a student who was working to finish their Ph.D. and had given a draft chapter of their dissertation to my colleague, their advisor, for review. Some of the data presented in the chapter looked odd and drew the suspicion of my colleague. He decided to take a look at the raw data files that are kept by the instrument that was used to take the measurements in question. Upon investigation he found that although the instrument files corresponded to some of the data presented in the chapter, large portions of the data had no corresponding files on the instrument. He suspected that some of the data presented in the chapter had been fabricated.

Apply the D.I.S.O.R.D.E.R. framework to this situation. The dilemma is defined as data fabrication. The advisor has already found some information about the extent of the data fabrication, but more information is needed about the potential ramifications associated with what the student has done. Begin by determining your institution's rules about research misconduct, in particular, data fabrication. Assume that the research was conduced with federal funding, choose an agency (such as the National Institutes of Health or the National Science Foundation) and determine what their relevant policies say. Continue through the D.I.S.O.R.D.E.R. framework by identify all of the stakeholders and exploring the options, being sure to pay attention to the rights of the student and rules of the institution and agency involved. Once you

[20]Office of Student Conduct and Community Standards, University of Wisconsin-Madison, "Academic Integrity," https://conduct.students.wisc.edu/academic-integrity/.

have all of this information, make a decision about what actions should be taken, then evaluate the effects that will likely be a consequence of that decision. You may find the that outcome is more severe or more lax than you expected. This is the time to review and reconsider so you can make sure that you have come to a decision that is reasonable. Repeat the O.R.D.E.R. portion of the framework and settle on a reasonable course of action.

5.7.2 OTHER RESOURCES ON RESEARCH ETHICS

Although this chapter touches on some key issues of research ethics, this is a broad field of study and one that entire books are devoted to. For additional reading on research ethics, the following references are suggested.

> National Academy of Sciences, N. A., 2009. *On Being a Scientist: A Guide to Responsible Conduct in Research*. National Academies Press (U.S.).

> Shamoo, A. E., and Resnik, D. B., 2009. *Responsible Conduct of Research*. Oxford University Press.

> Lipson, C., 2019. *Doing Honest Work in College: How to Prepare Citations, Avoid Plagiarism, and Achieve Real Academic Success*. University of Chicago Press.

5.8 SAFETY

From the basic ergonomics of your work environment to the issues that arise from the need to use dangerous substances, safety is a critical issue in your research environment. Even if you are not conducting experimental work yourself, it is likely that there are labs down the hall, on the next floor, or someplace in the buildings you frequent. Make sure that you know how to respond when something dangerous occurs that may impact your own safety and the safety of others.

Basic building safety features are the first thing to become acquainted with. You should determine where the emergency exits are located, where to go in the case of a fire, how to respond in the case of a natural disaster, and what actions your campus recommends in the case of an active shooter. Pay attention to drills and take part in training opportunities.

Be observant about the spaces that you inhabit—read the signage and identify safety equipment that is readily available (e.g., fire extinguishers, automated external defibrillators, emergency sowers, eye wash stations, etc.). You may not be using chemicals, operation lasers, or interacting with radiation sources, but this might be a regular activity in a lab down the hall, so be sure to note any signage about hazards near your work environment. Identify the location of phones for emergency calls—your cell phone may be most accessible but if a land line is readily available, use it. The location of the call can be more easily identified.

In the realm of experimental research, the potential dangers are myriad: chemical splashes, particle inhalation, burns from heat sources, etc. If your research is conducted in a laboratory or field environment, you will be required to take part in standard training associated with the common safety hazards and you will likely receive specific training on procedures you will use so that you understand and can handle the specific safety hazards. Training is essential and should occur BEFORE you begin in the research so that you know how to protect yourself and others, prevent hazardous situations from arising, and respond appropriately if a safety issue occurs. If you are not offered training, you should request it. If for some reason it is not readily available, seek out the information and educate yourself.

One of the keys to keeping a safe work environment is to ensure that the entire environment is a safe one, not just the area and materials that are directly your responsibility. If you see something, say something. If you notice another person in the lab doing something that puts them or others in danger, talk to them immediately. If you see a way to improve safety around a piece of equipment or standard procedure, talk to your research mentor or the laboratory manager. You should also make sure you know who to call if you see an unexpected problem, e.g., water leaking from under the door across the hall, or a noxious smell emanating from another laboratory. On most campuses each building will have a building manager or safety officer. Labs will have an emergency contact sheet on the door that also includes contact information. But, if in doubt, call emergency services using 911.

There are even health concerns sitting at your desk. You may spend much of your research time in front of a computer—coding, collecting data, analyzing data, and writing. It is important to have a setup that is ergonomic so that you do not develop issues over time—such as back or neck pain, carpel tunnel syndrome, etc. Maintaining the health of your eyes is also important. Eye strain is also something to guard against if your time in front of a screen is lengthy.

ASSIGNMENT 5-9:
INDIVIDUAL ASSIGNMENT – YOUR OWN SAFETY

Determine the three main safety issues relevant to your daily workspace. Investigate them in more detail to determine: What protections have been put in place to mitigate the safety hazards? What are your responsibilities with relationship to these safety issues? How can safety be improved and what actions can you take to suggest these improvements or make these improvements yourself?

ASSIGNMENT 5-10:
INDIVIDUAL ASSIGNMENT – CASE STUDY

Instructions:

Read the brief case description provided. Reread while noting the important information, and questions that are raised in your mind about the information provided, the individuals involved, and their situation. Determine both the basic issues and any deeper underlying issues at play. Consider the questions posed at the end of the case and how you would respond to these questions as well as other questions that could be asked of this case. Write a one-page response that includes a brief summary of the case and its issues, your answer to the questions posed, and recommendations based on your understanding of the situation posed in the case.

Case description:

Mary has done an excellent job in navigating the safety issues associated with her research project and is recognized in her research group for being adept with the logistics of handling both the day-to-day safety issues and the associated campus requirements. She is also a student member of a newly formed safety committee in her department which meets several hours every month. Her work on the committee has been helpful to her research group because Mary makes sure that they all stay up to date on the safety issues relevant to their work.

Dr. Smith, her research mentor, has recently taken on a new project and has already indicated that Jonah, a first year graduate student, will use this project for his thesis. The project is a very interesting one, but it will involve some new safety requirements and consultation with campus safety experts before it can be started. However, instead of having Jonah coordinate the safety requirements of the new project, Dr. Smith has asked Mary to take the lead.

Questions to consider:

Is it reasonable for Mary to take on this duty?

Is it possible for Mary to say "No" in the situation?

Will Jonah's lack of involvement have the potential to compromise safety?

How might Mary work with Jonah to get safety issues of the new project coordinated without overloading herself?

CHAPTER 6

Documenting Your Research Findings

6.1 KEEPING A RESEARCH NOTEBOOK

Several decades ago, when I worked in the medical device industry, the engineers spent the end of each day signing and dating their lab notebook pages and proving witness signatures as a cognizant individual on the notebooks of other engineers. When your notebook was full you turned it back to the company librarian and were assigned a new one. If you needed to reference one of your old notebooks you could check it back out again. These procedures were in place for data management and patent protection. Your clever ideas could result in patentable work and the signed, witnessed, and dated pages of your lab notebook might be used to prove that you were the first to come up with the invention. In 2013 a new patent law change went into effect in the U.S., moving patent priority from first-to-invent to first-to-file, however well-kept research documentation is still critically important today.

These lab notebooks, or research notebooks as I will call them, are an important tool for every researcher, whether an experimentalist, computationalist, or theoretician. The research notebook provides a place for you to document your thinking, your results, and your conclusions. These days your notebook make take the traditional form of a bound paper laboratory notebook or it may be an electronic document (or some hybrid form). If you have joined a research group, find out the practices of the group. The researcher in charge of the project (i.e., the principal investigator, or PI) may provide you with a physical notebook or give you an account to an electronic notebook. If not, it may be up to you to decide what format works best for you.

> **Student Perspective**
> "I work mainly on the computer and did not really understand how I could possibly keep track of anything outside of the digital medium. After looking at examples of lab notebooks in class and discussing what makes a good lab notebook I realized that my thought process and various other things to organize my thoughts and results could find a place in my notebook."

Ultimately though, it is common practice that the notebook will stay with the research group when you move on to your next position. This may be a requirement of the funding agency supporting the research on which you are working or an aspect of a broader data management plan of the research group or institution. You may be allowed to keep a copy for yourself (for instance a scan or photocopy of a paper notebook or a duplicate copy of an electronic notebook). The obvious exception to this would be if you are working on a classified project or working for a company where the intellectual property is owned by the company.

Although you are likely the primary person who will read this notebook other than your research mentor, keep in mind that it needs to be readable by others and this must be kept neatly and completely. Others should be able to reproduce your work based on what they read in your notebook. For instance, there might a student who follows on in your research area after you have left the research group. So, you should keep in mind that you are not the only person you are writing for. Your research notebook should be clear and understandable to someone working in the area.

The basic content of your notebook should do all of the following in order to create a traceable record of research progress/findings in one place where it can be easily accessed.

- Describe your research goal(s).

- Identify methods used.

- Support methods chosen with literature references.

- Include original raw data/images (or references to e-data).

- Include procedures/designs/programs/calculations (or references to e-files).

- Include final results and their interpretation.

- Attach print screens of e-data file directory hierarchies where applicable.

- Provide documentation that others can follow with enough detail that they could recreate the work.

- Describe thought processes, hypotheses, and outcomes.

- Plan future research activities.

- Write out steps to possible solutions for problems encountered.

Beyond the research itself, your notebook is a good place to record other research-related interactions and information. It is an excellent place to keep notes from lab meeting discussions, agendas for meetings with your research mentor along with the comments/suggestions made during those meetings, and a summary of research seminars that you have attended. Being able to refer back to these additional notes at a later time will become invaluable.

> **Student Perspective**
>
> "The longer that I continue to do research, the more pertinent that it is to have a good notebook, as I find myself looking back at certain past experiments and trying to evaluate where we've been, thus helping determine a plan forward."

You can also use your research notebook as a project management tool. Minimally it should describe the research goal or hypothesis you are currently testing. You can also use it as a roadmap for what comes next. This can be particularly valuable if there is a time lag and you will not be able to get back to your research to make further progress right away.

> **Student Perspective**
>
> "Documenting research properly and storing files in a way that makes sense has also been a skill I've had to develop over the past year. My lab notebook entries now are more helpful than they were at the start of the project. At some point I started ending each one with a list of immediate next steps, and I have found that really helpful, especially during the school year when I can't work on the project every day and need my notebook to remind me where I left off. I have gotten better about naming and storing files in a way that makes them easy to find again, which is important because as the project has gone on, I've collected a lot of files."

It may seem like a lot of work to keep a good research notebook and it is. However, doing so will pay off in a multitude of ways over time. Sometimes it is possible to make keeping a good research notebook faster and easier by creating procedures that you write once and then refer to (noting any modifications that you make over time) or making fill-in-the-blank tables for things that you do routinely.

Before starting your research notebook for a mentored project, talk to your research mentor. Your mentor may have expectations about what type of notebook is kept and what information must be kept in it. Additionally, if a federal agency or foundation is funding the research project, they may have requirements of their own (for example, the Nuclear Energy University Program funded by the Department of Energy put out a document titled "Proper Use and Maintenance of Laboratory Notebooks" with expectations for all the funded projects).

6.1.1   DOCUMENTING YOUR RESEARCH IN A PAPER LABORATORY NOTEBOOK

The lab notebook in its paper form has been used for hundreds of years. There are good reasons for this, other than calamities like a fire, they are long-lived documents. For example, we still have Leonardo Da Vinci's notebooks to look at today! Paper continues to be the way to record information for many.

Some basic expectations for a paper notebook are as follows.

- Use a permanently bound notebook with numbered pages.

- Ensure researcher name, contact information, and research group are prominently visible.

- Develop a table of contents as research is documented.

- Include a key to abbreviations used and naming conventions of samples/files.

- Date and sign each page.

- Write neatly in pen with lined through corrections; X any skipped/unused pages.

- Secure all additions with tape and signed/dated over edges (no loose pages, no staples).

Plan ahead with a numbering system for each notebook so you can reference between them. For instance my lab notebooks at the University of Wisconsin–Madison started with UWL1. If you are working on multiple large projects it may work better to devote different notebooks to different projects.

Hopefully you have developed some practice though laboratory classes you have taken in high school or as an undergraduate, but a research notebook is more than just the entries about experimental procedures/protocols and results. The research notebook is often the key place where everything gets tied together. For an experimentalist, this includes the reason why you are conducting the experiment, details about or references to protocols/procedures used, names of raw data files and output files from analysis, a plot of the results to date, the name of the folder containing relevant image files, and methods gleaned from a literature citation.

Many computationalists keep notebooks in addition to commenting their code so they can track their broader thinking about their research beyond the changes in functionality of a component of the code, along with version number of the code or output file name. Theorists keep notebooks to track their thought process, identify where ideas are built off of a literature citation, and capture their evolution in thinking about a topic.

In addition to notes about the research you undertake, you should also capture information about meetings you take part in, the seminars you attend, and key journal articles you read.

It may seem onerous at first, but the more you capture in you research notebook along the way, the easier your later work will be. You will find that a well-kept research notebook is particularly helpful in writing a paper or thesis. You will thank yourself later for developing good documentation habits early!

You will also want to regularly back up your research notebook. If it is kept on paper, this simply means making a photocopy or scan of the pages every month (or more frequently). For an electronic notebook you can export file or make an electronic duplicate that is kept on a different server (or alternate storage device) and in a different building. You have to think about the worst-case scenario—what if there is a fire in the building, or what if you backpack is stolen—it would be bad enough to lose a month of data but horrible to lose it all. Unfortunately, there are actual instances of this happening. Make sure you are not the star of the next cautionary tale!

6.1.2 DOCUMENTING YOUR RESEARCH IN AN ELECTRONIC RESEARCH NOTEBOOK

Some campuses, research institutions, and companies provide access to electronic lab notebook software. In some cases the use of a specific electronic lab notebook software may be required. These products are reasonably new and have varied levels of adoptions in different places. There are also a variety of different styles, including blogs, wikis, note taking software, and document management systems. As people have become more comfortable taking notes directly on an electronic device, electronic lab notebook products have seen increasing adoption. The advanced search functions and data management capabilities make these software options very attractive. Some of this software can also provide you with added ease in connecting from a variety of devices over the Internet and sharing with other research group members and collaborators. There can also be added benefits in electronic signing, file versioning, and activity tracking.

As with all software however, there are some lingering concerns over long term accessibility with software changes or a software company no longer providing updates to make the product compatible with new operating systems. Data security can also be a concern. Look into the software that your campus recommends or supports.

As discussed above, before you decide on how you will record your research activity, check with your research mentor about the practices and requirements of the research group.

6.1.3 REGULAR EVALUATION OF YOUR RESEARCH NOTEBOOK

Checkups are important and will help you to maintain a good level of completeness with your documentation of research. Some funding agencies will go so far as to require the researcher in charge of the project (i.e., the principal investigator, or PI) to regularly review your research notebook. Ideally this should take place regardless of such requirements. However, even if your research mentor is not doing regular reviews, it is good practice to periodically review it yourself. Use the activities below to do so. You can also ask your research mentor for guidance, by asking them to review your research notebook and provide you with feedback.

ASSIGNMENT 6-1:
INDIVIDUAL ASSIGNMENT – SELF-EVALUATION OF YOUR RESEARCH NOTEBOOK

Assess the last several months of your research documentation with the rubric below.

Check your PAPER research notebook for the following items:

☐ Name and contact information on the beginning of the notebook.

☐ Date and initial each page.

☐ Write in pen; cross out mistakes (but leave them legible); do not erase; do not tear out pages.

☐ Write neatly (so anyone can read it); leave space between things—do not crowd.

☐ No blank pages between entries.

☐ No loose pages; tape additions to a page.

Check your ELECTRONIC research notebook for the following items:

☐ Name and contact information on the beginning of the notebook.

☐ Logical naming system for each entry/file.

☐ Date and name on each entry/file.

☐ Electronic lock (i.e., archiving) activated on past entries/files.

Consider the following best practices for documenting research. Check the:

☐ Recording thoughts and ideas consistently.

☐ Statement of objective and description of specific work to be performed, or reference to an approved planning document or implementing document that addresses those topics.

☐ Identification of method(s) and computer software used.

☐ Identification of any samples, test equipment, and characterization equipment used.

☐ Description of the work as it was performed and results obtained, including names of individuals performing the work, and dated initials or signature, as appropriate, of other individuals making the entries.

☐ Methods and procedures described in detail and updated as needed.

☐ Description of any problems encounter and their resolution described.

☐ Entries clear enough so that the ideas can be reconstructed at a later date if necessary.

☐ Sufficient detail provided to retrace the investigations and confirm the results or repeat the investigation and achieve comparable results independent of the individual investigator.

What are you doing well? What needs improvement? Describe what action steps you will take in the next month to improve your documentation of research.

ASSIGNMENT 6-2:
INDIVIDUAL ASSIGNMENT – USING YOUR RESEARCH NOTEBOOK

Give an example of how you have used information that was previously recorded in your research notebook or in someone else's research notebook. How did you find that prior work? How was it helpful to you? Were you able to save time by having access to good documentation of prior work in a research notebook?

6.2   DATA STORAGE AND BACKUP

What would happen if your laptop was stolen or your hard drive crashed and the data was unrecoverable? Is that the only location your files are stored? If so, the thought should send you into a panic and make you immediately seek a method for backing up your data.

In research your data is not just yours. You are responsible for the data and its loss could negatively impact not just you, but also your research mentor, any peers and collaborators you are working with, and the scientific community as a whole. Funding agencies recognize this and many require a Data Management Plan to be developed and submitted with the proposal for funding. This Data Management Plan will include a description of the types of data to be collected (even file format types in some cases) and how that data will be managed and preserved. Not only backup systems to avoid any data loss but also how data will be shared with other researchers. The National Science Foundation expects the following items to be addressed[1]:

[1]National Science Foundation, "Grant Proposal Guide," Chapter II.C.2.j, https://www.nsf.gov/pubs/policydocs/pappguide/nsf15001/gpg_2.jsp#IIC2j.

1. the types of data, samples, physical collections, software, curriculum materials, and other materials to be produced in the course of the project;

2. the standards to be used for data and metadata format and content (where existing standards are absent or deemed inadequate, this should be documented along with any proposed solutions or remedies);

3. policies for access and sharing including provisions for appropriate protection of privacy, confidentiality, security, intellectual property, or other rights or requirements;

4. policies and provisions for re-use, re-distribution, and the production of derivatives; and

5. plans for archiving data, samples, and other research products, and for preservation of access to them.

The research you are engaged with may have a Data Management Plan in place and/or standards within the research group for collecting, organizing, storing, and backing up data. You should begin by asking your research mentor if there are data standards for your research that you should follow. However, some groups leave much of the detail up to each individual researcher, so in that case you need to think about some of these key aspects yourself.

Always keep the original is a basic rule. If you take an image you should keep an untouched original and make modification only to copies of the image file. It's a good idea to lock the original file so you don't accidentally make changes to it. In some cases a publisher may want you to submit both the original image file along with the version of the image that you will be putting into the figure. The same rule applies to raw data output from an instrument or codes. Keep the original output and make a copy on which you then conduct analysis.

You will obviously need your files somewhere you can easily access and work on them. This may be a laptop computer for instance. But you will also want to make sure ALL of your files are regularly backed up somewhere. This may be an external drive or cloud storage for instance. The data management professionals suggest that you have your data in three separate storage media in at least two separate locations in case one of them gets corrupted. This may mean that you have an external drive at home where you back up your files regularly and additionally you use a cloud data storage system like Box or Dropbox to keep a copy of your files in a separate location. Ideally these backups occur automatically without you having to manually initiate it, but if not, you need to get into a backup routine so that the time span between backups in minimized. Think of the worst-case scenario like your apartment building catching fire in the middle of the night—you get out safely, but your laptop and external drive are destroyed. However, if you still have a recent backup in the cloud then your months or years of research work are not lost.

When considering cloud data storage systems look into what is available through your institution. It is likely free to you and the institution has negotiated a license agreement that takes into account security and issues surrounding sensitive. It is also likely that you can easily share data with your research collaborators and reassign ownership of the folder to your research mentor when you graduate and leave the institution.

   Before you have too many files to deal with, step back and decide how you could best provide organization for your data. You will want to develop a logical folder system so that all the files are not jumbled into one place. They should also be named clearly and succinctly. Keep in mind that this collection of files will likely be accessed by someone else—maybe after you graduate—in order to extend and build upon your research. You need the structure to be logical and intuitive. A text README.txt file at the top of the file structure can help you to describe how things are organized and keep a list of key data attributes that will be helpful not only to other but to yourself. Here is a simple file structure to illustrate these ideas:

```
Creep Crack Growth
    README.txt
    Equipment
        Drawings
        Images
        Quotes
        Validation
    Experiments
        CCGProcedures
        CCGTesting
            Alloy617
                Analysis
                CrackGrowthData
                Microscopy
            Alloy800H
                Analysis
                CrackGrowthData
                Microscopy
        HeatTreatCharacterization
            Analysis
            RawData
    Publications
        ExpMech-TechniquesPaper
            Drafts
            Figures
            FinalSubmission
        MetMatTransE-CCGPaper
            Drafts
            Figures
```

                         FinalSubmission
                    SEM-ConfProc
               Reports
                    AnnualReports
                    FinalReport

Having an agreed upon naming convention for experiments or version control system for software is especially important when you are working collaboratively. If each part of the name for the folder/file is defined and everyone has this shared information then a sample with the name

U4-H9P63Y98-PS12-PS20-L30-PDMS5M-D4-1Hz8V-D19

will tell everyone what it was part of experiment **U4** that involved a specific protocol of that number, and that the cell type used was **H9 p**assage **63** with a **y**ield purity of **98**%. The sample was **p**re-**s**eeded on day **12** which ended on day **20** when the cells were put on **l**anes of **30** micron width using a **PDMS** substrate with Young's modulus of **5** kPa and an extracellular matrix of **M**atrigel. Pacing was then begun on day **4** in the lanes at **1 Hz** frequency at **8 v**olts and ended on **day 19**. This experiment has a page-long README file that provides detailed information about the naming system so every researcher on the project knows what is happening at each stage of the experiment and can add to the naming string as appropriate when they take data after doing the next step of the protocol with the samples.

Software version control allows you to track how software changes over time and keep old archival versions to return to as needed. Versioning is also critical so that you know what results were created by which version of code and that multiple people can work on a code simultaneously. Platforms like GitHub allow individuals and groups to deposit their code. Similar to the README file, a wiki within the shared resource provides people with information about the code, how to use it, etc. This would be overarching information beyond what would already be included in the comments within the code.

> **Student Perspective**
> "I've collected a lot of files. I started using GitHub for my code, not because I have to share it with anyone, but because I find it's helpful for keeping track of what version is the most current, and I think it's a good idea to have a backup copy of it somewhere other than my laptop hard drive. I should be better about keeping other particularly important files in the cloud as well, so if my laptop were to die, it wouldn't be as big of a problem (I do keep my laptop backed up using an external hard drive, so my files would not be permanently gone, but in the short term, files there wouldn't be as easily accessible as files stored somewhere in the cloud)."

There a number of resources available on best practices for data management.[2] Your campus will likely have an office of Research Data Services or librarians who can help you with campus-specific resources and practices.

6.3   AVOIDING DATA MANIPULATION

We may not often think about it, but there are some important ethics issues when it comes to handling data, figures, and images. As discussed in the previous chapter, this begins by avoiding error, negligence, and misconduct. As discussed above, the first step is to collect and retain all original output, raw data, and image files. The original version should be locked and left untouched. A copy should be made when the file is needed for further analysis.

Obviously, fabricating data is wrong, but sometimes it is more fuzzy when considering falsification. For instance, you may have what you believe to be outlier data points. If there is a documented reason (for instance a comment in your lab notebook about high room temperature due to broken HVAC in the building, a sample contamination, or a power fluctuation during data collection), then it is reasonable to exclude those data points. However, if you do not have a known reason for exclusion, you will need to report all the data points. In some cases, you may be able to show statistically that the outliers do not fall within the data population, but even in this case you would report any statistical exclusions that you made when writing about your results.

There are also subtle ethics of presenting data in best light vs. manipulating the presentation of the data to make a false impression. For instance, you may see a small trend in your data ranging between 90 and 100 that is interesting. However, if presented on an axis scaled between 80 and 100 this will exaggerate the trend to a reader who is not scrutinizing the graph carefully. When you present your data, you want to do so in a way that is honest and discloses the full picture. It's fine to focus in on the small trend, just be sure to do so in a way that is not misleading.

With the advent of digital images and the capability to manipulate them, a few ethical lapses in data manipulation have been made very public. Because of this, some journals regularly screen submitted images to guard against image manipulation and publish standards of practice.[3] There are some general guidelines that you can follow when handling images with software packages, like PhotoShop and ImageJ, that will keep you on the side of good ethics.[4] To start with, keep an archival copy of every image you generate that you never manipulate. If you need to crop or change contrast for example, then, work on a copy of the original image and log every change that you make to that image either in your research notebook or a text file stored

---

[2]See for instance, DataONE. A collaborative project on data management funded by the National Science Foundation (NSF). https://www.dataone.org/.

[3]"Image integrity and standards," Nature Research, Springer Nature Limited. https://www.nature.com/nature-research/editorial-policies/image-integrity.

[4]Hendrickson, M., 2010. "Digital Images," a talk presented in "Optical Microscopy Course," W. M. Keck Laboratory for Biomedical Imaging, University of Wisconsin-Madison.

with the image(s). If you have made any manipulations beyond simple cropping and changes in brightness or contrast, then describe what you have done when you present the image. This can be described in the figure caption or the methods section of your paper. You should avoid things like modification of a part of an image, aggressive cropping of an image, using extreme or nonlinear adjustments in intensity, and digital filtering of an image. Furthermore, you should always present representative examples of the results you have observed. If a particular image was an outlier, then it must be described as such if you want to present it.

CHAPTER 7

Sharing Your Research via Oral Communication

7.1 INFORMAL CONVERSATIONS WITH OTHER RESEARCHERS

This chapter discusses a variety of different ways in which researchers need to communicate their work orally. It is an important skill to develop so that you can comfortably talk about your research with a range of different audiences, at different levels of technical depth, and with different levels of formality.

> **Student Perspective**
> "I was very surprised that the researchers have to care a lot about communicating with different groups of people, like the general public, reporters, students, colleague, and etc. A stereotype of a mad scientist who does not communicate with other people at all is actually not possible in reality."

One of the most important audiences with which you will need to communicate are the other researchers in your field. These may be people you work with every day, a research mentor you communicate with regularly, collaborators you communicate with periodically, or other engineers and scientists that you interact with at meetings and conferences. It may be tempting to shy away from these sorts of communication initially, but they are truly critical both to your personal development as a researcher and to the research project you are undertaking. Even if it pushes you outside your comfort zone a bit, you need to engage in these conversations.

> **Student Perspective**
> "Overall, I have learned that perseverance, confidence, and communication are the most important skills that a researcher can possess. In order to have a successful project, one must communicate with other scientists to resolve problems in an appropriate and timely manner as well as be able to resolve issues when help is not available."

ASSIGNMENT 7-1:
INDIVIDUAL ASSIGNMENT – RESEARCH MEETING UPDATE

Each research group has its own style and practices involving meetings, but at some point you can expect that you will need to talk about your work in front of others in your research group. This may be more or less formal and may or may not involve preparing presentation slides.

Your assignment is to present short research talk of 5–8 minutes in duration. The talk must be relevant to your research topic but may range in content from a review of a paper to an explanation of some aspect of your ongoing research. Formal slides are not required but you may use them if it is customary in your research group.

7.2   INFORMAL CONVERSATIONS WITH NONSPECIALIST AUDIENCES

It is frequently the case that we have the need or desire to talk about our research with people who are not specialists in the field or even comfortable with technical subject matter. It may be important for us to convey the purpose of our specific research project and/or the motivation behind the general area in which we are doing research. Communicating with people outside your research group might originate from a need to write a cover letter to a journal editor about a manuscript you are submitting, discuss your research with a program manager or funder of your research group, prepare for a job interview, or communicate with the general public about policy decisions related to your research.[1]

Certainly, developing these non-specialist communication skills are to your own advantage when it comes to getting your research funded and published or landing a new job when you complete your degree, but these skills are also important because of their societal benefit. Your ability and willingness to explain your research has the benefit of promoting a scientifically literate society, so that people who have an opportunity to influence the paths of future research and technological uses are doing so with a basic understanding of the related research. Scientists and engineers also have an obligation to report back to the taxpayers who fund their work.

To do this type of communication effectively you need to invest time in figuring out what it is you want to convey and the best way to go about doing it. Depending on the specific audience you will have to change the amount of depth and technical detail you discuss, eliminate the jargon you might normally use with colleagues, provide more background about the subject area, and shift your emphasis from details of project to a more general discussion of its potential applications.

When you engage with other researchers in your area you have the opportunity to use jargon to speed up the transfer of information in conversations. However, you need to be sure that

[1]Baron, N., 2010. *Escape from the Ivory Tower: A Guide to Making your Science Matter*. Island Press.

you know what this jargon means to others so that you don't have a problematic miscommunication. Some junior researchers try to hide their limited understanding behind jargon—this may seem to work, but if you really don't know what you are taking about the improper use of jargon will soon give you away. When you communicate with nonspecialists you need to remove the jargon, and this forces you to really understand what's underneath these words.

The order in which you talk about things will also need to change. The big idea, or headline, needs to come at the beginning to keep the listener engaged. Narrow the key points that you will make to only those that are essential for the particular audience you have in mind. Include some relevant and memorable facts or theories related to each key takeaway message. When you actually talk to people, don't be afraid to repeat yourself—the main message should be touched on multiple times throughout.

Several proponents of informal science communication focus on story telling as a way to convey research concepts to general audiences.[2] This can be particularly effective when communicating to public audiences. Traditionally, stories that people tell have a hero and a goal. The "hero" can be a person (you) or a thing (like a technology) and the goal should be something we care about (such as an outcome that will make people's lives better). Like any good story, there are obstacles along the way that the hero must overcome to succeed. As you tell the story, your goal is to engage the listener so they want to know what happens next. However, success is not a requirement for every story... sometimes it is a "tragedy"... for example, the equipment broke! Even these sorts of stories are important ones to tell, it's one of the realities of research that people don't often understand.

Although you may not want to craft an entire talk to fit a traditional narrative story arc, an anecdote can help to "seduce" your audience into paying attention. The key is to make sure that your anecdote offers a concrete example that is representative of the research.[3]

ASSIGNMENT 7-2:
INDIVIDUAL ASSIGNMENT – HONING YOUR MESSAGE

Identify a topic area—either a current research project or a topic related to your research interests—and craft talking points that you would want to convey if you had five minutes to convince someone to fund this type of research. Don't think of this as preparing a speech, but rather as preparing for a conversation in which you want to make some key points and be able to respond to anticipated questions. Develop a topical "headline" and 3–4 main messages with 1–2 pieces of supporting evidence. Develop a brief closing summary statement that links back to the headline captures themes brought up in your main message.

[2]Miller, T., 2015. *Muse of Fire: Storytelling and The Art of Science Communication*, https://www.spokenscience.com/publications/. Olson, R., 2018. *Don't be such a Scientist: Talking Substance in an Age of Style*. Island Press.
[3]Laszlo, P., 2007. *Communicating Science: A Practical Guide*, Springer Science & Business Media.

> Headline
> Main Message 1
>   Supporting Evidence
> Main Message 2
>   Supporting Evidence
> Main Message 3
>   Supporting Evidence
> Closing Summary Statement

For example, below is a brief example of once main message on the topic of graduate education:

> Main Message: Graduate education is integral to university-wide goals
>   Evidence: Grad students are essential in research and undergrad education
>   Evidence: Grad students increase the diversity of the campus community

> Relevant data would also be identified on the number of research and teaching assistantships that graduate students hold and the national and international diversity statistics of the graduate student population at the institution.

ASSIGNMENT 7-3:
INDIVIDUAL ASSIGNMENT – MESSAGE BOX

You can use the Message Box[4] shown below as a graphical method (Figure 7.1) to aid in honing your message. The Message Box is a communication tool developed by Nancy Baron, who has worked in science outreach with Seaweb and COMPASS.[5] It was created for scientist so that they can prepare for interactions with the media and policy makers, but it is a generally applicable communication tool to organize the main points surrounding a technical topic in preparation for a discussion with a non-expert.

The Message Box helps you to create "talking points," or key points that you believe are important to cover in a conversation about your topic. You can use this framework to explain what you do to those who know little about your area of expertise. It is a flexible tool that can be used not only to prepare for a verbal conversation but also for written communication, such as a cover letter, press release, or website.

The Message Box itself is a tool for you to use to organize your thinking. Using a piece of paper or a PowerPoint slide, divide the page into four quadrants with the *Issue* in the middle. Create a list of 2–4 talking points around the *Problem, So What, Solutions, and Benefit*. Although

---

[4]Baron, N. and Weiss, K., 2007. "The message box" preparation for talking with the media, seminar given on September 21.

[5]Baron, N. and Weiss, K., 2007. "The message box" preparation for talking with the media, seminar given on September 21.

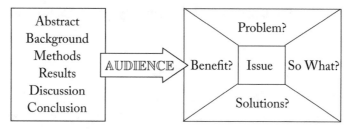

Figure 7.1: Technical presentation of information (left) vs. Message Box (right). Consider your audience when you translate into Message Box talking points.

it might be helpful to show it to your research mentor for feedback, it is not something you would show to someone you are speaking with about your research. It is a tool for your to facilitate your communication.

ASSIGNMENT 7-4:
INDIVIDUAL ASSIGNMENT – VIDEO COMMUNICATION PRACTICUM

It is often useful to have practice presenting in front of a camera so you can see how you sound and critique your own presentation. This assignment will give you an opportunity to gain experience talking in front of a video camera in a low risk setting. Find a place where you can be undisturbed and use your phone or computer to capture the video.

There are a number of questions you can choose to address during your video session. Ideally, you should focus on 1–2 questions because the total time of the video should be approximately five minutes. Although you should prepare some talking points in advance, it is best if you do not read off of notes.

Questions to consider.

Why should someone consider (or not) entering your major?

Why should someone consider (or not) going to graduate school?

What have you learned about the process of research?

What strategies would you suggest for finding a research project and research mentor?

What tips would you give someone just beginning their research project?

What tips would you give to someone about staying on track with research progress?

What advice would you give to someone preparing an oral presentation?

Review the video after you have recorded it. Identify things you have done well and things you would like to improve.

7.3   ENGINEERING OUTREACH

One specific type of nonspecial communication comes in the form of "outreach"—those events that campuses and communities hold to engage the public with science and engineering topics. Often these are targeted to the K-12 age group and frequently involve some type of hands-on activities. Even if your campus does not hold an Engineering Expo type of event, there are frequently opportunities for undergraduate and graduate students to interact with kids about engineering through K-12 schools and after school programs.

Ideally, you pick a topic to present that you are interested in, maybe even something related to your research. Engaging more of the senses will make their experience memorable. It is particularly helpful if people can get their hands into and onto an experiment or demonstration materials. You may be able to help a group experienced with outreach who already has activities developed. If not, there are numerous resources available which can provide you with content that you can use and adapt. If the professional society in your discipline does not already have materials available, try looking at places like `TryEngineering.org` or `volunteer.asee.org`.

Although precocious middle and high school students who voluntarily come to campus for a science or engineering outreach event can surprise you with the level of their knowledge about technical topics, many people understand less about basic science concepts than you might expect. For example, "four out of five Americans do not understand the concept of a scientific study sufficiently well enough to provide a short sentence or two of explanation.[6]" Because of this you need to think about ways to engage with people to understand what they already know before you begin an explanation. You can do this more readily by asking questions and interacting with them about the topic area. Ideally, you do this in a fun an engaging way, rather than making the person feel like they are taking a quiz.

> **Student Perspective** Student Perspective
> "In short, I learned it is better to teach someone something simple well than it is to teach them something more complicated badly."

When presenting, it is important not to overload people with too much information. You don't want to talk down to them, but you need to simplify the ideas that you are trying to get across without introducing errors or creating misconceptions. Often you must adjust as you are interacting with people based on how they are responding. It's like running an ongoing

---

[6]Knight-Williams, V., Santigian, L., and Williams, D., 2005. An overview and annotated bibliography of the use of analogies, hands-on activities, models, simulations, and metaphors: Part III of front-end analysis in support of nanoscale informal science education network. (Knight–Williams Research Communications.).

experiment! You may have to try different things to see what's most effective and be willing to modify your original plans.

> **Student Perspective**
>   "[Doing outreach] turned out to be a good practice exercise for me, because I hadn't anticipated that most of the people that came to the exhibit were people that had never taken a chemistry class. So for the first hour or so, I found myself struggling to keep people interested in the topic simply because I was out of practice explaining things that I wrongly assumed were common knowledge. For example, many people didn't understand atoms bond to one another. This is actually a fairly complex concept for people who haven't had a science heavy education like me. Through this, I learned the miracle of teaching through visual aids. The molecular model kits that we had were really useful for those complex yet basic concepts that I needed to explain, especially when I was speaking to younger patrons. I learned that I really need to improve on how I present technical topics to younger people and people who are less interested in science in general. Part of it is a matter of using the right amount of scientific detail, but a lot of it has to do with how engaging I can make the topic seem as a presenter."

Below are some basic tips to think about before you interact with the public about a science our engineering topic in an informal science education setting.[7]

- Know the intended audience.

- Define a limited set of learning goals (2–3 at most).

- Be aware of the length and attention span of the audience.

- Use multiple modalities to address a range of learning styles.

- Don't assume prior knowledge.

- Define terms and avoid jargon.

- Avoid graphs, especially multidimensional graphs and log scales.

- Explain what you see in scientific images and diagrams.

- Use metaphors and analogies that explain and enlighten.

- Include personal aspects of the story, not just the scientific facts.

[7]Crone, W. C., 2006. Bringing nano to the public: A collaboration opportunity for researchers and museums, S. E. Koch, Ed., *Nanoscale Informal Science Education Network*, Science Museum of Minnesota, St. Paul, MN.

- Repeat the message, explaining it in multiple ways, but be concise.

- Provide clear directions for an activity.

- Encourage visitor conversation.

- Test for misconceptions.

- Evaluate at every stage.

The arts can also be used to engage public audiences with engineering topics and employed as an entry point for those who might not be as intrinsically interested in science and engineering topics. For example, in collaboration with professional artists and science museum exhibit designers I have worked with other engineers and scientists to engage audiences by using silver and gold nanoparticles suspended in polymers to make "stained glass" artwork with the public. It's amazing what creative minds can come up with!

The most important thing I learned when working with museum exhibit developers was that you have to make it fun. If it is not fun people can just walk away or walk on by. This gives you an opportunity to be creative! For example, I worked with a balloon artist several years ago to develop an interactive balloon model for a carbon nanotube structure that we could build with kids visiting outreach events. It was a huge success and the activity is now being used by outreach presenters all over the world.

Hopefully you will engage in outreach yourself at some point if you have not already done so. Being the presenter can be fun and it will likely remind you of why you got excited about your area of study in the first place.

ASSIGNMENT 7-5:
INDIVIDUAL ASSIGNMENT – EXPLANATION FOR AN 8-YEAR-OLD

- Pick a topic that you will explain to a group of 8-year-olds (e.g., rainbow, fluorescent light bulb, hibernation).

- Develop a 30-second explanation (oral, written, movement, and/or illustration).

- Be prepared to share your explanation.

7.4   POSTER PRESENTATIONS

The research poster is a common form of communication, both on campus and in poster sessions held at research conferences. The poster size is usually designated but is often somewhere between 3′ × 4′ and 4′ × 8′; a size large enough to enable viewing by someone standing a few feet away. In some cases, the poster content and organization are prescribed, but more commonly they simply follow the general organization of a research article. The poster title, authors, and their affiliations usually appear across the top, with the title in a larger font (72–100 point font is common for a title). The content sections of the poster usually feature an abstract, background, methods, results, and conclusion (the body text is usually 24–32 point font, with larger font for section headings). References and acknowledgments are usually at the bottom of the poster, often below the conclusion. The layout should progress logically from left to right and top to bottom.

Above all, a poster should be designed to be visually appealing, with graphics, figures, and images as a key focus and large in size. The overall design and color scheme should be harmonious. You can use boxes and borders to set apart different sections of the content. Although it is important to include some text, it should be carefully chosen to be both informative and brief. Black or very dark text on a white background is easy to read. You may want to think of the poster as an advertisement or announcement of your work—thus "It needs good copy.[8]" Make "headlines" and easily readable text to go along with great visuals. In many circumstances, you will accompany the poster to provide a verbal explanation.

> **Student Perspective**
> "Through the creation of a [conference] poster… I learned much about scientific communication. First, many basic skills were cultivated through this poster designing process such as making something readily understandable, uses of bullets highlights and bold, use of white space, and knowing your own poster. Second, I learned that the number one priority when creating a poster is, making it simple to understand for the audience and aiding in breaking down any barriers which might have confused yourself. In addition, the honing of the elevator speech was a process. There was a large amount learned on what not to say (difficult or misleading concepts) and what to say (most widely appealing/relevant virtues of my work)."

When preparing a poster, it is critical to determine the requirements before you begin. If it will be presented in a conference venue, then the conference organization will provide guidelines. It is essential that you find out the size restrictions and whether the poster will be displayed in a portrait or landscape format. It is also important to determine what printing process you will be using so that you can find out if there are restrictions on the size, how much whitespace should be left around the edges, and how much advance time will be needed to get the poster printed. You do not want to print out the poster only to have the last three inches cut off, forcing you to both revise your poster in a hurry and spend money to print it twice. Carefully proofread your poster before it is printed. Also determine how you will transport the poster—with some

---

[8]Laszlo, P., 2007. *Communicating Science: A Practical Guide*, Springer Science & Business Media.

printing processes you can fold the material and pack it in a suitcase, but if it is printed on paper you will need a poster tube (be sure to label the tube with your name, address and phone number).

You must also consider the verbal part of the poster session. You need to have your talking points thought out in advance and designed to augment the poster content. As with non-specialist audiences you need to engage in a conversation and be prepared to discuss your work at the appropriate level of detail for the people you are discussing it with. This means that you might actually have two (or more) levels of explanation based on whether you are talking to a specialist who works in an area close to your own or someone who is interested in the work but not an expert in your technical specialty. Practice what you will say out loud before the big day. Ideally, you will do this with your research mentor and/or members of your research group who can give you feedback on both the poster content and your accompanying explanation.

During the poster session, be open to a conversation as you are getting across your main points. Ideally, a poster session is an opportunity to engage in discussions that will help you to move your research to the next stage and you may even have great questions posed by non-experts that help you to think about your research in new ways. If you are nearing a junction in your academic career (i.e., applying to graduate schools or job hunting), you should let that be known in the conversation. You can even bring along business cards and copies of your resume in case an appropriate opportunity arises.

7.5   THE RESEARCH TALK

Presenting your research may feel overwhelming at first, but it will get easier the more times you do it, both in terms of the time it takes to prepare your talk and how comfortable you will feel in giving the talk in front of an audience.

When preparing a research talk, the first thing to determine is who your audience will be. Your research mentor should be able to tell you the kinds of people who are expected to attend. This will help you understand the level of jargon that can be used and how thoroughly you will need to discuss the background for your topic. The second thing to know is the time limit on the talk. In many cases the amount of time you will have to present will be quite ridged. Conference talk slots are usually somewhere in the range of 12–18 minutes, whereas a seminar talk could be 50 minutes. Be certain to ask if the time you are being given is inclusive of questions so that you know whether you need to make your talk a bit shorter to allow time for questions. With these two pieces of information you will be able to determine the scope of your talk. You will likely need to pick and choose what you talk about from the research you have done—don't try to cram everything in.

Develop an outline for your talk and select the key visuals that will accompany it. Often the flow of a talk is similar to a paper with background, methods, results/discussion, and conclusion. If you have already submitted an abstract to a conference then your topic will be somewhat fixed, but you may need to do more thinking about how you will motivate the work and what background you need to provide so the audience understands where your research fits into the field. Showing that you know the context of your research and have read the related literature is an important aspect of the talk and will help to convince the audience of the knowledge you have developed. As with writing, your presentation will also need to give credit to others where appropriate. This means that you will include the names and/or references to the work of other researchers/groups on your slides (as well as in the words that you say). In addition to a title slide at the beginning that may include information about co-authors and funding, you will also include an acknowledgments slide at the end. Be certain to include your research group, other collaborators, and all of the funding that supported the research.

Emphasize the visuals in your talk and add a few key phrases or bullet points on each slide. You will use your spoken words to fill in the details. Having less text will also prevent you from falling in the trap of simply reading the slides to the audience. You can use figures and images from the literature as long as you give credit to the original source. This is best done with the citation information on the slide where it appears (rather than a number and a reference list at the end). You will also be presenting figures, graphs, images, and/or videos about your results, but you may find that you need to supplement these with additional images, animations, and graphics that fill in the gaps and visually explain your approach and methods.

Discuss your outline or draft slides with your research mentor early so that you can determine if you are on the right track. Once you have developed a complete draft of the talk, practice it independently and in front of others. Many research groups will have practice talks in their regular group meeting as a conference approaches, but if this is not planned then ask your research mentor and others in your research group if they would be willing to watch a practice talk and give you feedback. Be sure to organize this far enough in advance so that you can make changes and implement the feedback thoroughly. Practice the revised talk again before the big day.

When you speak, do so facing the audience. If you plan to use a pointer, practice using it so that you hand is steady. Speak with a volume that will be heard by everyone in the room. If you will have a microphone, determine how to turn it on an off and check what position will pick up your voice clearly. As you give your talk be sure to actually explain your slides to the audience. Although you are very familiar with your research, your audience will need to be oriented so that they can understand them too. This means that you will need to take time to point out the axes on a graph, the color coding for a figure, or the definition of the symbols that appear in equations.

During your practice session, be certain to ask for questions that you might get from the real audience. This will help you to practice thinking on your feet and constructing some of the responses you might use. For obvious questions that you just don't have time to address in detail in the main talk, you might prepare a few backup slides that you can place at the end of your talk and refer to as needed. During your actual presentation, it may happen that you get a question you don't know how to respond to. Don't panic. You can reply gracefully with "Thank you for that question. I'm not certain of the answer at the moment, but that is something I will look into." If you get a more aggressive question that calls the underlying basis of your work into question and you are not certain how to deal with it, you can reply "That is a much broader discussion than we have time for now, but I would be happy to talk with you in more detail after the session." Of course, then you will need to be certain to track down this person after the session and have that difficult discussion, but at least you can have the discussion without a large audience.

> **Student Perspective**
> "Another thing that my research project has helped me improve on has been my presentation skills. I've given short presentations about my project, or topics closely related to it, a number of different times…. I've gotten better at putting slides together that are clear and informative, and at judging how much material I should have for a presentation that is to last a specified amount of time. I've also gotten better at explaining my project in a way that makes sense to people outside the … field. Increased familiarity with the field has helped me be better able to explain it as well. Practice presenting to people is also always helpful for improving my presentation skills. Every time I present I am less nervous about it, and I think in general my presentations have gotten better over time."

Whether you plan to give your talk from your own laptop, submit the file in advance to conference organizers, have it on a file sharing system, or external memory device, be certain to have it available in more than one place just in case the primary source does not work for some reason. Some conferences have a speaker prep room set aside where you can check to make sure your talk is working correctly and run through it one last time. Make sure that any videos or animations you are have included are functioning properly. If possible, especially for something like an oral proposal talk or thesis defense talk, practice the talk in the room you will be using so that you can be one familiar and comfortable with the space. At a bare minimum, make sure that the technology will work in advance of your talk.

Whether the talk will be given in a class or a conference, arrive early so you won't feel rushed. Make sure the host, instructor, or session chair knows that you are there. Ask if there are any timing signals that they plan to provide. If not, you can ask a colleague to give you a discreet wave when you are within a minute of the end time. Running long is considered rude so you should avoid doing so.

Take a deep calming breath, release it in a slow steady exhale, and then give a great talk!

ASSIGNMENT 7-6:
INDIVIDUAL ASSIGNMENT – THE FLASH TALK

Sometimes it is actually easier to give a long talk than a short talk, but in some circumstances you will be given a tight time limit. In a recent conference I attended, the graduate students either presented their research in either a "flash talk" or a traditional poster session. The flash talk had a strict 3 minute time limit—exactly 180 seconds. These short format presentations vary, but they are designed to give a large number of speakers time to share a glimpse of their work, usually prior to some time frame in which people will be able to mingle and follow up with speakers whose research interested them.

Essentially this talk is an advertisement for your work, so you want to get across a few key items to encourage follow-up. Be sure to include your name, institution, and email address. After mentioning the topic, which should be succinctly summarized by the title of your talk, begin with a brief motivation for the work. Discuss the method of research you are using to approach the problem. Present a summary of your most important results. Sum up with conclusions you have been able to draw from the work you have done. In some cases you may be talking about work in progress. If so, the first half of the talk stays the same. If you have preliminary results to share, you can include those. You will wrap up with your plans for future work.

Hone your slides to get across key visual information with your talk. There are several strategies for doing this effectively. One strategy is to use a single slide layout (see quad chart format below). This gives the speaker a chance to talk without worrying about changing slides on the computer and allows the audience to look through all the information over the 3-minute time frame. Alternatively you can break up the information in a traditional slide format—title slide, motivation, methods, results, conclusions—no more than 5 slides can be reasonably covered in 3 minutes. Ideally you should condense it to fewer if possible (for instance: title and motivation, methods, results, conclusions).

Once you have the draft talk prepared, do a practice run and time yourself. It is likely you will have to adjust both the slides and what you say. Practice several times to get your talk to be exactly 180 seconds. You may want to use an audio recorder so you can listen to your talk and make adjustments after listening to it critically. You may also want to practice in front of someone. You may find that you tend to talk faster or slower in front of an audience and it is important to know this in advance.

ASSIGNMENT 7-7:
INDIVIDUAL ASSIGNMENT – THE QUAD CHART

Another term you will sometimes hear regarding short presentations of research is the "quad chart." Sometimes this is intended to be a standalone entity and other times it is used as a backdrop to a short presentation. It is somewhat reminiscent of a poster in that all of the information is contained in one view. It is often created using PowerPoint and usually intended as something that is projected on a screen all at once or printed on a single page. Funding agency program managers may request a quad chart be produced for a research project that is underway, or it might be used to summarize a completed project.

It might seem a bit reminiscent of the Message Box discussed earlier but the Message Box is a tool for you to use and is not shown to someone you are speaking with, whereas the quad chart is the main product being delivered. Although there is some overlap in the content between a quad chart and the Message Box, the audience and intent is usually quite different.

There is no single format for a quad chart other than that it is usually broken into four quadrants. The quad chart is intended for a technical audience and it shares many features with a poster. It also has the formality of a title, authors (e.g., research project participants), and funding acknowledgments.

Usually specific instructions are given by the requester (often a funding agency program manager) about what is to be included. One common format is to use the following content in the quadrants: (1) an image or graphic that depicts the overall project; (2) a statement about the objective(s) of the research; (3) a description of the approach being taken to reach the objective(s); and (4) a timeline that lists milestones and progress to date.

When engaged in a Department of Energy research project several years ago, we were asked to provide a quad chart for a progress report meeting that involved the principal investigators (PIs) of all the research projects being funded by the program. The instructions were very explicit about what exactly was to be included in each quadrant, but it basically boiled down to the following key elements.[9]

> **1st quadrant:**
>
> **Purpose/Objective**—"A short description of the major contribution envisioned from the project."
>
> **2nd quadrant:**
>
> **Importance/Relevance**—"Highlight the relevance of the research being conducted."

[9]Nuclear Energy University Program, Department of Energy, https://neup.inl.gov.

Title Authors/Project Researchers Institution(s)/Affiliation(s)	
Purpose/Objective:	**Importance/Relevance:**
Tools/Methods/Facilities:	**Sample Results:** **Status of Deliverables:**
Citations/References: Acknowledgments/Funding:	

Figure 7.2: Sample quad chart format.

Impact Areas—"Identify the areas of impact from the successful completion of the project. These should provide a broad view of the project's scope as opposed to the more technical 'Purpose/Objective' section."

3rd quadrant:

Tools/Methods/Facilities—"Include the details of the various specialized set of tools/methods and facilities that are being used/developed as part of the project."

4th quadrant:

Sample Results—"Highlight the key result obtained from the work done to date. A figure should be included along with a caption that explains the key findings."

Status of Deliverables—"Include the list of the deliverables submitted in the project proposal and indicate the status of each deliverable."

For this individual assignment, use the quad chart format in Figure 7.2 to summarize the status of the project you are currently working on.

ASSIGNMENT 7-8:
INDIVIDUAL ASSIGNMENT – CASE STUDY

Instructions:

Read the brief case description provided. Reread while noting the important information, and questions that are raised in your mind about the information provided, the individuals involved, and their situation. Determine both the basic issues and any deeper underlying issues at play. Consider the questions posed at the end of the case and how you would respond to these questions as well as other questions that could be asked of this case. Write a one-page response that includes a brief summary of the case and its issues, your answer to the questions posed, and recommendations based on your understanding of the situation posed in the case.

Case description:

Fan has been a graduate student in the Hoffman research group for four years. Her progress as a graduate student is going very well. She has successfully passed her qualifying exams and preliminary exam and her research project has been producing excellent results. However, she's beginning to feel invisible and worries that no one recognizes her research accomplishments.

In the weekly research group meeting, Prof. Hoffman always asks for a volunteer to give a more detailed presentation in the following week. Fan faithfully gives her brief research update, but she never volunteers to give the more detailed presentation. People frequently ask her to repeat herself when she gives her updates, which makes her self-conscious about her English proficiency.

This week in the research group meeting several other graduate students in the research group are presenting draft abstracts for the upcoming annual conference in the field, but Fan's advisor did not ask her to prepare one even though Fan thinks her research is ready. In previous years when other students have come back from the conference, she hears about the interesting talks they attended, how well received their own talk was, and senses that they come back even more energized about their research projects. Fan is beginning to get worried that she will never get a chance to go to a conference and present her research.

Fan is friendly with Susan, another graduate student in the group who started at the same time as her. She asks Susan to help her to make a case with Prof. Hoffman about giving her the change to attend the annual conference.

Questions to consider:

What might Susan suggest that Fan do to get an opportunity to attend the conference?

What other assistance can Fan seek from campus resources to improve her presentation skills?

7.6   RESOURCES ON ORAL COMMUNICATION

Although this chapter touches on some key issues related to oral communication, this is a broad topic that entire courses and books are devoted to. For additional content, the following references are suggested.

Baron, N., 2010. *Escape from the Ivory Tower: A Guide to Making your Science Matter.* Island Press.

Hayes, R. and Grossman, D., 2006. *A Scientist's Guide to Talking with the Media: Practical Advice from the Union of Concerned Scientists.* Rutgers University Press.

Humphrey, J. D. and Holmes, J. W., 2008. Style and ethics of communication in science and engineering. *Synthesis Lectures on Engineering*, 3(1):1–140.

Miller, T., 2015. Muse of Fire: Storytelling and The Art of Science Communication, `https://www.spokenscience.com/publications/`

Olson, R., 2018. *Don't be Such a Scientist: Talking Substance in an Age of Style.* Island Press.

Vernon, B., 1993. *Communicating in Science: Writing a Scientific Paper and Speaking at Scientific Meetings.* Cambridge University Press.

CHAPTER 8

Sharing your Research via Written Communication

Although writing might not be the first thing you think of when you imagined the sorts of things you would do as an engineer, it is an essential aspect of nearly every engineering position and a skill that you can develop to maximize your career outcomes. In engineering research, the ability to share your methods and findings with others via written communication is essential.

8.1 TRANSLATING TECHNICAL TOPICS IN WRITTEN FORMATS

Before delving into technical writing, it is helpful to begin by translating technical topics into more general explanations. This will allow you to use your prior writing experience and help you hone your explanation skills. Much of this approach echoes the early sections of the previous chapter. Your goal in writing about technical topics for nonspecialists is to translate the technical so that it is more broadly understandable. Initially, this can also be a very helpful way for you to develop a deeper understanding of a new research area you are engaging with.

In the long run you will also need to be able to translate your own research so that broader audiences can understand what you have done or what you plan to do. This comes up in a variety of contexts, but commonly it ties into funding your research. Working in industry you would need to be able to write a memo about your research/development work so that your boss' boss can understand its importance and how it impacts the company and its products. Working in an academic institution often requires communicating with the public through press releases and updates to both alumni and donors. Many funding agencies also require you to write a "lay abstract" about your research proposals and often expect short research updates that are understandable for public consumption.

As discussed earlier, you will need to tailor the depth and technical detail depending on the specific audience you want to communicate with. You will also need to provide a more general discussion of the research, avoid the details, and focus on the potential applications. In some cases, you will have the benefit of working with a professional communications specialist, but often we are left to our own devices to figure this out. One way to be successful is to see how others have done it so you can emulate their approach. The experts in this area are science writers, and you can find their work in print and online. Look for good science writing in places like the *Science* section of the *New York Times*, *National Geographic Magazine*, *Wired*, DiscoverMagazine.com, ScientificAmerican.com, and the news written about research at your own institution by the university communications writers.

ASSIGNMENT 8-1:
INDIVIDUAL ASSIGNMENT – LABORATORY-TO-POPULAR-PRESS[1]

Current scientific research is often covered by the popular press. What happens to a scientific idea as it travels from the lab to the newspaper, news blog, or web magazine? How is scientific information "translated" by the press for the general public? Is press coverage accurate, objective, and complete?

Look for recent media coverage of research in your area of interest. Sources might include the business/technology section of a print newspaper, popular science magazines, web magazines, or science news blogs (for example: DiscoverMagazine.com, ScientificAmerican.com, EurekAlert.com Spectrum.IEEE.org). After finding an article of interest, use your literature search skills to find the peer-reviewed journal paper related to this article that has been published by the researcher(s) in the scientific literature.

Write a 2-page paper about the original research and the media coverage. Begin your paper with a brief summary of the research and the results based on the journal article. After this summary, critically consider mass media reporting of the research described in the journal article. What aspect of the research was emphasized? Was anything important omitted? Were the results accepted uncritically? Were conflicting opinions discussed?

ASSIGNMENT 8-2:
INDIVIDUAL ASSIGNMENT – SEMINAR PRESS RELEASE

Attend a research seminar and write a summary in the style of a short "press release" for a general audience. Summarize the seminar talk in 250–500 words using the following structure.

- Include a short and enticing title.

- Use the first few sentences to introduce the speaker, their university affiliation, the date and title of the talk, and the seminar forum (e.g., the Mechanics Seminar Series at UW-Madison).

---

[1] Adapted from Caitilyn Allen, Department of Plant Pathology, University of Wisconsin–Madison.

- A short description of the main finding(s) and relevance of the work should appear early in the summary.

- The remainder of the "press release" should provide additional information about the findings presented, including context for the work that was presented, how the work advances the field through new advances, new methodologies, or reinforcement of prior work.

- Minimize the amount of jargon you use and if you must include a technical term be certain to explain what it means.

- Do not write about the miniscule details for the research presented and use active voice.

- Include a quote from the speaker if appropriate.

8.2   BASIC PRINCIPLES OF TECHNICAL WRITING

Technical writing, which would include writing of a thesis, technical report, or engineering journal article, is different in structure, tone, and format from other types of writing.

Initially, it may be challenging and awkward for you to write in this style. Two styles of the same events are described below—the first in a "normal" description that I might write down as a description of something that happened in my day in a diary, the second in a style more appropriate for a technical journal.

> I came home from work and was greeted at the door by my chatty cat, Marty. Her insistent meowing made me realize that her food bowl was empty. After setting down my backpack, I filled her bowl and she immediately began to eat, inhaling half of what I had fed her.

> At 6:20 pm, a vocalization from feline subject #1 (female) was noted upon initial interaction. Within one minute, 0.5 cups of Cat Chow (Indoor Dry Cat Food, Purina) was dispensed into the feeding bowl positioned at ground level. The subject commenced ingestion within 5 seconds, and 0.3 cups was consumed by 6:34 pm.

Occasionally students will question why they must write in a particular style. It is a good question to pose, because certainly information can be effectively communicated with a variety of writing styles. However, you will notice in the simple examples above there is more specific information available. If I gave you the second example and asked you to take care of my cat because I was unexpectedly called away from home, you would know that my cat expects to eat a little after 6 pm, how much to feed her, where to place the bowl, and what kind of cat food to buy if none was left. An engineering researcher must be able to master technical communication techniques to provide the detail necessary to fully describe their research so that it can be replicated and do it in such a way that their work will be accepted and acknowledged by the field.

Every discipline has its standards and forms that it uses. You must know how to play by these rules before you can consider bending or breaking them. For example, if you were an aspiring screenwriter trying to launch your career in Hollywood, you will want to write it in the customary format so that your ideas are presented in a familiar way and the producer does not discard it as amateurish before even reading it. Once you have hit the big time, and written several blockbusters, you can choose to write in a different style, but the likelihood is that by then you will have discovered why the standard style has evolved for this discipline and writing in that style will no longer feel foreign to you. Whether you are an aspiring screenwriter or engineer the style in which the people in that field communicate will seem foreign at first, but ultimately you will learn how to write in that style effectively and it will become more natural.

8.2.1 DEALING WITH WRITER'S BLOCK

Whether a seasoned writer or a novice, nearly everyone gets stuck at some point and finds it hard to either start writing or make progress on writing. Before launching into the particulars of specific types of technical writing, it is useful to have some strategies to employ if you run into a snag with the writing itself. Here are some suggestions that you might try.

- You don't have to start at the beginning, and usually the abstract is what gets written last anyway. Try starting with the section that you find easiest first so that you can gain some momentum.

- Make appointments with yourself for writing. This is very useful when you have a large writing project that you need to accomplish over a period of time. At the time you have designated on your calendar, you must write, no excuses.

- Create the diagrams/charts/figures you believe tell the story of your work, put them in a logical order, and then go about describing them. Describe the methods used to acquire the data found in your figures, what the reader should see when they look at the figure, and what conclusions can be drawn from the figure. This text will likely end up in different sections of the paper (Methods, Results, Discussion), but sometimes it is easier to write about a figure and move the text that you generate to the appropriate section at a later time.

- Some people find that developing a progressively more detailed outline is a fruitful strategy. In this case you would begin with a skeleton outline, then add detailed bullet points to it until you can eventually turn the bullets into sentences and the sentences into paragraphs, ultimately fleshing out each section of the paper.

- Try "free writing" where you just capture your stream of consciousness. This means that you don't edit along the way or search for just the right word. You don't even worry about sentence structure and punctuation, you just get the ideas captured. Later you can go back to the product of your free writing and begin editing and sculpting these ideas into the appropriate format.

- Talk to a friend about your work and record the conversation. Encourage them to ask you probing questions about your research. Afterward, review the audio file and type the pieces that seem useful, adding more to the verbal description as you are transcribing it.

8.3   STANDARD FORMATS IN TECHNICAL WRITING

There are several basic mechanisms of communications in the world of research, and in written communications there are a number of standard formats that are used. Writing can be used for very different purposes, so it is important to understand both the purpose, and the audience for whom you are writing. Initially your writing may be intended solely for your research mentor. As time progresses, however, you may begin writing for venues in which other researchers in your field will be the primary audience.

Who your audience is will have an impact on a variety of different aspects of your writing. When writing to be read by members of your research community you can use more technical jargon, however, it is critical that you understand this jargon rather than hide behind it. Any missteps in the use of jargon will be quickly identified by experts in the field. When writing to a more general audience, or a broader technical audience who are not members of your specific subspecialty area, you will need to be much more limited in your use of technical terms and jargon. Choose a few of the most important technical terms, and make sure that you clearly define them through your writing. The layering on additional jargon should be avoided if at all possible.

8.3.1   ABSTRACTS

The abstract is likely the most ubiquitous format in technical writing. For instance, you may want to submit your work to a conference for a poster or oral presentation; in most cases this will require you to provide an abstract that summarizes the work you will present. Abstracts are also commonly used at the beginning of longer forms of writing such as technical reports, proposals, theses, and journal articles.

A good abstract will provide motivation for the work, information about the approach taken in the research, and a summary of the key findings. Ideally, as part of the abstract you will also indicate how the research findings impact the field. Commonly the abstract will include: the context for the research, a description of the methods used, the important results/findings, and the impact/importance of these results. Depending on the context, the length of the abstract may be prescribed. Commonly abstracts fall in the range of 200–300 words, although an "extended abstract" will be the longer in length. In general, it is not appropriate to include citations and abbreviations in an abstract.

Take some time to read abstracts in your discipline area that have been written for a similar purpose to your current writing task. For example, if you are writing a conference abstract, then look at example abstracts from the previous conference year. If you are writing an abstract for a journal article or research paper, then look at the abstracts of recent journal publications in your discipline. While reviewing these example abstracts, think about the type of information that is being conveyed by each sentence. Dissecting what others have included in their abstract will help you decide what readers will expect to see in yours.[2]

Some publications require what is referred to as a "structured abstract" which includes specific subheadings.[3] This began in fields of health and medicine to assist clinicians for quickly identifying methodology and results in journal articles, but this format is being adopted by more science and engineering journals because it helps the reader to quickly identify relevant content. Subheading may include some or all of the following: Introduction, Objectives, Methods, Results, Discussion, Conclusion. Each subheading will usually have 1–3 sentences of concisely written content that the reader can easily understand.

As a reader of journal articles we usually begin with the abstract, but as a writer of a manuscript to be submitted as a report or journal article you will generally find it easiest to write the abstract last (or at least after you have completed a substantial amount of the other writing involved in your manuscript). These sentences maybe the ones you spend the most time crafting in the entire manuscript!

ASSIGNMENT 8-3:
INDIVIDUAL ASSIGNMENT – ABSTRACT DISSECTION

Identify a journal paper of importance to your research area, preferably one that is considered to be an influential paper or has been highly cited. Begin by reading the abstract, then read the rest of the paper. Return to the abstract again and dissect each sentence by identifying how it addresses one or more of the four components below:

- Motivation/objectives

- Approach/methods

- Results/findings

- Impact/significance

---

[2]"Writing an Abstract for Your Research Paper," UW-Madison Writer's Handbook, The Writing Center, University of Wisconsin–Madison. https://writing.wisc.edu/handbook/assignments/writing-an-abstract-for-your-research-paper/.

[3]National Library of Medicine, "Structured Abstracts," https://www.nlm.nih.gov/bsd/policy/structured_abstractsabstracts.html.

8.3.2   REPORTS

The technical report is a common form of writing for engineers. This may be required within your research group as a standard mechanism of keeping your research mentor updated, or as an expectation of research funding. If you hold a fellowship you may be required to provide a report at the end of the fellowship year, for instance. Many federal agencies, private foundations, and industry contracts that provide funding to support research activity also expect regular reports on the progress being made. These reports may be expected monthly, quarterly, or annually. Although your particular research contributions may only be a portion of the reported activity, you will likely be responsible for providing not only data and results, but also a written summary of the work to date.

> **Student Perspective**
> "I had to write quarterly and annual reports for my portion of the project and was involved in the discussion of how we were going to proceed with the project for renewal in funding."

Before you begin your writing, determine the required format and expected detail. Often, your research mentor or other research group members will be able to provide you with this type of background information. As you embark on your writing, focus on presenting information in a logical order, using clear sentence structure. Consider whether figures and tables will assist you in presenting the information more effectively. Relevant figures might include a schematic diagram of a process, a photograph of an experimental setup, a graphical depiction of results, and/or a micrograph.

ASSIGNMENT 8-4:
INDIVIDUAL ASSIGNMENT – WRITTEN RESEARCH UPDATE

Some research mentors expect students to provide weekly or monthly written reports, but even if that is not the case for you on a regular basis there may be the need for you to occasionally provide a written research update (e.g., when travel prevents your usual face-to-face meeting). If your research mentor has asked for a specific format of update, you should provide feedback in that format. If a specific format is not required, then present your update in a logical manner broken down by project and objectives. Provide actions taken since the last meeting/report, results obtained, next steps planned, and questions that need to be addressed. In preparation for your next interaction with you research mentor, draft a written report.

8.3.3   TECHNICAL WRITING FOR A PROPOSAL, THESIS, OR JOURNAL ARTICLE

Entire courses and books are written on the topic of technical writing alone.[4,5,6] In this section we will focus on some key highlights and strategies you can use to help improve your writing. In addition to this content, you may also want to avail yourself of other resources such as textbooks on the subject and campus resources (e.g., your university's Writing Center).

For most formal writing, there are standard format requirements, or at least a few options for acceptable formats. Begin your writing process by understanding the format expectations. In the case of a thesis, there may be detailed guidelines provided by your program or institution on both the structure and layout out of the document. If you are writing a journal article, then you will need to consult the guide for authors that the journal publishes (usually on its web site). Depending on the journal they will be more or less prescriptive. Regardless, these author instructions should be followed to the letter. If the macro-scale organization of the document is not predetermined, then you will need to work with your research mentor to develop the basic outline. This will usually include: abstract, introduction, methods, results, discussion, conclusion, acknowledgments, references.

It is also the case that within the discipline a certain style off writing is expected. As discussed above, you will need to become aware of those stylistic expectations and to use them in your writing. This is oftentimes easiest to do by looking at examples of prior writing in that style. Remember that there are both good examples and bad examples available to you, so you may want to ask for advice about which examples are the ones you should use to guide your writing.

Whether it be a proposal, thesis, or journal article, you must convince the reader to invest their time in what you have written. Having it clearly written, well organized, and visually appealing is essential. You also have to anticipate that the reader may not actually read the document in the order that you have presented it. For example, if I am deciding whether to invest time in reading a journal article, I'll first read the abstract. If the abstract looks worthwhile, then I'll glance through figures and jump to the conclusion to see what the key findings were. If my interest is piqued, then I will invest more time in reading everything in between. For the actual writing process that you will undertake, you will actually write the abstract and conclusion section LAST because these get written after you know everything you want to say. You will likely spend far more time per word writing and rewriting these sections of the paper, so that they are as clearly and compellingly written as possible.

[4]Humphrey, J. D. and Holmes, J. W., 2008. Style and ethics of communication in science and engineering. *Synthesis Lectures on Engineering*, 3.1, 1–140.

[5]Northey, M. and McKibbin, J., 2009. *Making Sense: A Student's Guide to Research and Writing*. Oxford University Press.

[6]Day, R. A. and Gastel, B. *How to Write and Publish a Scientific Paper*. Cambridge University Press.

8.3.3.1   Persuasive Writing

Although it must always be scientific and objective, technical writing should also be considered persuasive writing. This is obvious in the case of a proposal where you are trying to pitch an idea and potentially convince someone to fund that idea, but it is also true for technical writing in general. You must always persuade your reader of the merits of your work with logical argument, compelling evidence, and engaging language. You may also need to consider addressing any counter arguments in order to deal with common objections preemptively.

Choose an introductory paragraph from a prior writing assignment that you have completed and rewrite it while focusing on persuasion. The original paragraph should be from a technical writing topic, although it does not have to be related to your research. Include both the original and revised paragraph in your assignment.

8.4   REFINING YOUR WRITING

Different people vary in their writing speed and effectiveness, but I have never met someone who was able to sit down and write the final version of their paper on the first try. As Paul Silvia, professor of psychology, puts it: "Writing productively is a skill, not a genetic gift, and you can learn how to do it.[7]"

Good writing requires reworking and editing what you have written (sometimes dozens of drafts are required). Additionally, when writing with the input of your research mentor or coauthors for a journal publication, you will need to seek their feedback and incorporate modifications. The process can sometimes be long and frustrating, but the outcome will be substantially improved.

> **Student Perspective**
>
> "I suppose I was under the assumption that good experiments and more so good researchers are those who can finish projects as quickly as possible. After what I know now, this seems to be true and false at the same time. On one side, researchers are undoubtedly judged by how many published papers they have produced, which is directly related to how quickly they can start and finish projects. But, on the other hand, they are also judged by the substance of their publications, which could be related to how *slowly* and carefully they start and finish projects. It seems that there is a very fine line between this race to produce as many published papers as you can and trying to maintain a standard of quality of information being produced."

Ideally, the process of writing will start well before the research has been completed. There are a number of ways to go about writing productively while you are conducting research.

---

[7]Silvia, P. J., 2007. *How to Write a Lot: A Practical Guide to Productive Academic Writing*. American Psychological Association.

- When you read an important journal article that you expect to cite, take some time to write about the key aspects of the article. You may be able to use the notes section of your citation management system to capture this type of writing.

- As the methodology that you are using to conduct your research becomes solidified, it is an opportune time to begin crafting your methods section.

- Writing about your results as they begin to accumulate allows you to organize this information more effectively and this practice can be helpful in identifying previously unanticipated gaps that you will need to fill.

It is important to work far enough ahead of your deadline to give enough time for you and the others invested in your writing to make the needed iterations. In the case of a thesis you may also be expected to provide a complete draft weeks in advance of your defense date.

Writing does not have to be done in isolation. You will likely have peers who are also in the process a writing (or should be) at the same time you are. Even if it is not helpful for you to write in the same physical space, it can be helpful to set writing goals that you hold each other to and provide each other with feedback periodically.

8.4.1   WRITING WORKSHOPS

In some course settings you may be asked to review the written work of a peer in your class and provide constructive feedback. Ideally you will be given some instruction and a rubric to do so. If not, the guidance in this section should be helpful to you. Keep in mind that the author will be reading your review, so you should take pains be constructive in your criticism and to state the positives with the negatives. Every author, whether a classmate or a seasoned researcher, is a person just like you and me! They will be more open to the criticism and making changes to their writing based on your comments if they are framed constructively.

When giving feedback you might ask the receiver if they want constructive criticism, and how it can be best delivered. This will show that you are willing to adjust how you provide the feedback and it will allow the recipient to reflect on how they can best receive it.

Bradley Hughes, Director of the University of Wisconsin–Madison Writing Center, helped me to develop guidelines for writing workshops that we hold twice a semester in our research course sequence. Over the years I have asked students for feedback on how to improve these guidelines for responding to writing on engineering research topics. The resulting suggestions below can be used in a writing workshop forum or when exchanging your writing with a peer for feedback.

Some Suggestions for Responding to a Colleague's Draft[8]

Before reading the draft–

1. Find out what the writer is intending to do in the document and who the intended audience is.

2. Find out what the writer wants from you as a reviewer at this stage of their writing and use that information to prioritize your feedback.

When reading and responding–

3. Read the entire draft before commenting.

4. Praise what works well in the draft; point to specific passages; explain why these passages work well. PICK AT LEAST ONE THING to compliment and begin your response with that.

5. Describe what you found to be the main point of the draft so that the author can determine if their intent has been achieved.

   (a) Try describing what you see in the draft.

   (b) What you see as the main point?

   (c) What you see as the organizational pattern?

6. When providing criticism, be honest (but polite and constructive) in your response. Try responding as a reader, in the first person (e.g., "I like _____." "I got lost here ..." "I think you could help readers follow this if _____").

7. Time is limited (for your response and for the author's revision), so concentrate on the most important ways the draft could be improved. Comment on large issues first. Consider the following questions.

   (a) Is the purpose of the document clear to a reader? Does the draft achieve its purpose?

      i. Is the writing accessible to a scientifically literate audience with some background in your area of research?

      ii. Are ideas presented in an interesting manner?

      iii. Can the reader infer what the specific aim of the research is? Are the goals clearly stated?

---

[8]Adapted with permission from Bradley Hughes with modifications and edits from Engineering Physics majors at UW-Madison. See "Peer Reviews," UW-Madison Writer's Handbook, The Writing Center, University of Wisconsin–Madison, https://writing.wisc.edu/handbook/process/peerreview/.

    iv. Is the scope of the project clear? What are the deliverables of the research?

    v. For a proposal:

        A. Does the writer propose the research is such a way that it appeals both to a general technical audience and members of the author's specific research field?

        B. Does the writing provide a compelling argument for the significance of the proposed research?

    vi. For a report, manuscript, or thesis:

        A. Is enough background given so the reader understands how the data was collected, or how a theory was developed?

        B. Is the draft convincing in its argument to support the conclusions? Are the results clearly documented? Is evidence used properly?

(b) Are ideas adequately developed?

    i. Are the important ideas of the work presented? Is there a clear focus?

    ii. Is the draft effectively organized? Is the sequence of points logical?

    iii. Is there an appropriate balance between major and minor points?

    iv. Do the author's ideas flow logically from one paragraph to the next?

    v. Were there any paragraphs within the author's draft that seemed out of place?

    vi. Are the transitions between sections strong? Is material from earlier in the document built upon and referred to clearly in later sections?

(c) Is prior published work on the topic described in sufficient detail to give context to the current work? Are the references clearly cited?

(d) Are the figures/tables/equations clear and appropriate?

    i. Do the figure captions provide appropriate detail?

    ii. Do the figures/tables support the claims that are made in the text?

    iii. Are the mathematical equations understandable?

8. Be specific in your response (explain where you get stuck, what you don't understand) and in your suggestions for revision. And as much as you can, explain why you're making particular suggestions.

9. Identify what's missing, what needs to be explained more fully. Also identify what can be cut.

10. Engage in a discussion, but refrain from arguing with the author or with other respondents.

11. Mark proofreading edits (awkward or confusing sentences, style, grammar, word choice, proofreading) on a printout to hand it to the author rather than spending time on these details in the discussion.

ASSIGNMENT 8-5:
GROUP ACTIVITY – WRITING WORKSHOP

If you do not already participated in a class or research group that holds Writing Workshops, you can form your own writing group with peers at your institution. A group of 4–6 people is ideal. You will need to agree on the frequency of your meetings, how long you will meet, and what deadlines you will pose on sharing your writing prior to the workshop. A suggested format follows.

Writing Workshop Group

You will each need to produce a piece of writing by midnight on Monday for the writing workshop you will be participating on Wednesday. Provide a copy of your written piece, including the cover page information discussed below, to all of the other workshop members. Before meeting, everyone must read the written pieces of all the other group members and come to the workshop prepared to discuss the writings. Bring a copy of your own written piece and cover page as well so that you can reference it and make notes.

Writing Assignment

Choose a research report, journal article manuscript, research proposal, or thesis you are working on as the subject of your writing piece. Provide 3–5 pages of new writing to your Writing Workshop group members. Figures and tables (if needed) as well as references should be attached to the end and should NOT be counted toward the 3–5 pages. Include an outline of the overall piece with a description of where this writing will be incorporated.

If you are not actively writing at this time, ask your research mentor to identify a "good" thesis in the same general field as your research topic. Read this thesis and write a 1-page reflection commenting on the organization of the thesis, what you learned about thesis writing through your reading of this "good" example, what was done well by the author, and what modifications you would suggest to improve the thesis.

Writing Workshop Cover Page

The following questions should be addressed in the cover page of the writing piece.

1. What part of your proposal/thesis is this draft (for example, the introduction to my thesis; or the review of technical literature; or the first part of the results section …)?

2. What are your *main* points in this section?

3. What *specifically* are you happy with and do you think is working well in this section?

4. What *specifically* would you especially like some feedback on or help with in this draft?

5. Anything else your readers should know to read this draft in a way that will be helpful to you?

ASSIGNMENT 8-6:
INDIVIDUAL ASSIGNMENT – WRITING WORKSHOP REFLECTION

Reflect on the Writing Workshop activity. Discuss the parts of the process that worked well and what could be improved. Consider "Suggestions for Responding to a Colleague's Draft" in Section 8.4.1 and how it can be refined for technical writing. What are specific critical questions that must be asked for the type of writing you reviewed? Should the questions differ when considering a journal article manuscript vs. a research proposal vs. a thesis?

8.5   ISSUES SURROUNDING AUTHORSHIP

Who is included as an author and the order of the authors can become a contentious subject because it involves both getting credit for the work and taking responsibility for the work. To avoid or at least minimize such problems, it can be helpful to talk about authorship when you are embarking on the research, well before you get to the stage of writing. As an early stage researcher, it is a natural topic for you to bring up for discussion with your research mentor so that you better understand the norms within your research area.

Shamoo and Resnick suggest beginning the determination of authorship by identifying the ways in which individuals have contributed to a research project. They identify the following areas of research contribution[9]:

- Defining problems

- Proposing hypotheses

- Summarizing background literature

- Designing experiments

- Developing methodology

- Collecting and recording data

- Providing data

- Managing data

[9]Shamoo, A. E. and Resnik, D. B., 2009. *Responsible Conduct of Research*. Oxford University Press.

- Analyzing data

- Interpreting results

- Assisting in technical aspects of research

- Assisting in logistical aspects of research

- Applying for grant/obtaining funding

- Drafting and editing manuscripts

Who appears on the author list can be more complex, particularly in a larger project that has involved a number of people at different stages of the work. In some cases, the journal will identify the criteria authorship. The medical community has spent time wrestling with this issue as a result of some historical problems where individuals were included on the author list although they did not contribute to the work. The International Committee of Medical Journal Editors (ICMJE) proposes the following criteria for inclusion as an author on a journal publication[10]:

- substantial contributions to the conception or design of the work; or the acquisition, analysis, or interpretation of data for the work; AND

- drafting the work or revising it critically for important intellectual content; AND

- final approval of the version to be published; AND

- agreement to be accountable for all aspects of the work in ensuring that questions related to the accuracy or integrity of any part of the work are appropriately investigated and resolved.

You will notice here that the last bullet explicitly deals with the authors taking responsibility for the work that is published. In large collaborative projects, and particularly as a junior colleague, it is difficult for you to know about the details of every aspect of the work. Certainly, you have responsibility for the aspects of the research and writing that you were directly involved with, thus you can ensure that those parts are conducted in the most ethical manner possible. And, if for some reason the publication is called into question, you have the responsibility to provide information related to the research.

Disciplines and sub-disciplines have different ways of determining author order: whose name goes first on the author list, whose name goes last, and in what order others appear. In some disciplines, it is simply alphabetical. In many disciplines, the principal investigator of a research project is usually the last author. The student or researcher who conducted the majority

---

[10]International Committee of Medical Journal Editors (ICMJE), 2019. Defining the Role of Authors and Contributors, http://www.icmje.org/recommendations/browse/roles-and-responsibilities/defining-the-role-of-authors-and-contributors.html#two.

work and wrote the majority of the paper is usually the first author. A paper that coincides with a chapter or more of the student's thesis will usually list student's name first, with other individuals such as the research mentor on the author list.

ASSIGNMENT 8-7:
INDIVIDUAL ASSIGNMENT – AUTHOR ORDER IN YOUR DISCIPLINE

Identify three journal articles published by your research group and look at the individuals in the author list. Using the author affiliation information given in the journal article and your personal knowledge about your research group, identify the roles that each author holds in research group or in other collaborating research groups (e.g., undergraduate student, graduate student, postdoctoral researcher, scientist, principal investigator, etc.).

Make an appointment with your research mentor to discuss the common practices of authorship in your discipline. Using the journal articles that you have identified, discuss author order on these examples in the context of common practice within your discipline.

8.6   PUBLISHING YOUR RESEARCH

The first question to consider is whether or not your research can be published in a journal. In some cases, research is done that can't be published openly or its publication has to be delayed for some period of time (often referred to as an embargo period). These publication restrictions usually only come up if you are working on classified research or working under a non-disclosure agreement. As a student this is not a desirable circumstance because you need to publish your work to build your resume. If you believe this may be the case with your research, it is important to talk to you research mentor about what aspects of the work will be publishable, and how you will be able to build your credentials so that you are ready for the job market when you complete your degree.

For the majority of work conducted at university campuses, external presentation and publication restrictions are seldom an issue. However, you and your research mentor may decide to delay dissemination of your work because of a desire to patent. If this is the case, you will be working with your campus research office to determine the patentability and submit a patent application. They will help you determine the appropriate timing the public disclosure of your research (e.g., a conference presentation or journal article submission).

Aside from the cases above, journal publication is the primary outcome of the engineering research that you will do (as well as conference proceedings publications in some fields). This allows other researchers, and people interested in the field, to learn from and use your findings. Adding to the body of knowledge in the open literature helps everyone to move the field forward, and often enables future advances in technology and products that benefit society.

The point at which the research is ready for publication is a judgment call that your research mentor will help you determine. But often, there is a desire to get your work published sooner rather than later, especially if you are working in a fast moving and competitive field of research. You would also prefer to have publications listed on your resume when you apply for a job, so it is in your best interest to help get the research completed and the manuscript submitted for publication. However, in the end, it will be up to your research mentor (or the principal investigator of the project) to make the determination of when the research is ready for dissemination.

> **Student Perspective**
>   "My previous understanding about publication of research was that is important but not essential. I thought that getting published was not a *necessary* condition for career advancement. I thought that other methods of disseminating information like conferences, colloquia, and informal group meetings and conversations with other institutions were of equal importance to being published. I assumed that conferences were the best way of spreading ideas, since those ideas are being told by the originator with the opportunity for immediate questions and/or feedback. Conferences are important, but the most important way of spreading information and ideas is through journal publications."

Publishers use similar review processes for evaluating manuscripts that they receive. The schematic[11] in Figure 8.1 gives the general flow of the decision-making process—both from the journal's perspective and the options you and your co-authors have once a decision has been rendered. Your research mentor will likely provide guidance on both choosing an appropriate journal to submit your work to, and the details of how to go about submission.

It is essential to keep in mind that for a coauthored paper everyone must agree on the final version prior to it being submit it.

Although the review process can seem adversarial, the ultimate goal is to assure that the research being published has been rigorously conducted, well documented, and written about in a clear manner. Usually the comments that come back from a reviewer will help you to improve the writing and clarify what you have done and the conclusions you have drawn from your outcomes. Sometimes the review may identify additional work that should be completed prior to publication, e.g., an additional control experiment or a validation run that was missing. Other times the reviewer may ask for something that is out of scope of this manuscript or it may seem that the reviewer does not understand the fundamentals of the work you are doing. When this happens, it is possible to write a rebuttal to the editor asking that a particular review or portions of the review be set aside. You will need to work with your coauthors to determine the best

[11] Adapted from: Barker, k., 2006. *At the Bench: A Laboratory Navigator*, Updated Edition, Cold Spring Harbor Laboratory Press, Cold Spring Harbor, NY.

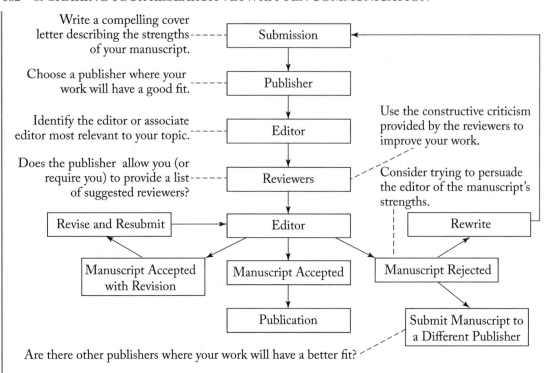

Figure 8.1: Flow of the review and decision-making process in taking a manuscript to journal publication.

course of action once you receive your reviews. It is important to act quickly though, often the response to reviews must be submitted within a deadline period.

> **Student Perspective**
> "The process of getting published involves a fairly rigorous (when done correctly, anyway) peer review process. The data is scrutinized, the ideas analyzed, and conclusions examined before the information is ever released to the scientific community. This generally prevents bad data and poor science from being published, thus preventing wasted time and funds by other scientists attempting to build on others' work."

Because the process is rigorous, depending on many people doing a variety of difficult tasks, and involves multiple levels of communication, the publication process is time consuming. Doing the research and writing the manuscript are certainly the majority of the work and time spent, but completing the manuscript submission, responding to reviews, and making revisions will take weeks or months. You need to be prepared for this additional work.

> **Student Perspective**
> "I was very surprised to find out how long it takes for a journal to accept and publish an article. After submitting an article for publication, it can take months to hear back on whether your article was accepted or not. Then, if your article does get accepted, it can take even longer for it to actually be published. The longest wait I was able to find when looking through papers this semester was around 13 months, which might have been *the* most surprising thing I learned all semester. But, with a little more searching I found that many journals are starting to post accepted articles online before their actual publication in the journal. I think this is a step in the right direction, as it will definitely help get published articles to the community faster..."

For nearly every journal you will need to do some writing beyond the manuscript itself. The guide to authors published on the journal's website will detail the additional items required for submission. Often a cover letter to the editor is expected—the journal may prescribe the contents, but it often is expected to include information about the importance of your findings, how your work fits into the scope of the journal, the most appropriate associate editor to handle your manuscript, and assurances that the manuscript is not under consideration with another journal. The response to reviewer stage of the process will also require writing—usually this includes a letter that only the editor sees, as well as a detailed written accounting of how you are responding to each of the reviewers' points that is usually seen by both the editors and reviewers. Because the reviewers will also have access to this written response to the reviews, it is critical to be respectful and use carefully crafted language when you are in disagreement with a reviewer's point.

ASSIGNMENT 8-8:
INDIVIDUAL ASSIGNMENT – WHERE TO PUBLISH

Identify potential journals where you might publish the research you are currently working on. Begin by looking at the papers that you are currently citing, and journals that they are published in. Also consider other key journals in your research area that you may be familiar with, or that your research mentor has mentioned. Determine whether your topic area is a good fit for aims and scope for these journals. Identify the Impact Factor of these journals and other relevant statistics provided, such as the timeframe between submission and publication. Look at each of the journal's web pages and identify the information/guide for authors.

After you have considered several journals, identify the top three candidates and summarize why you think these journals would be a good fit for your research.

8.7   RESOURCES ON WRITTEN COMMUNICATION

Although this chapter touches on some key issues related to written communication, this is a broad topic that entire courses and books are devoted to. For additional content, the following references are suggested.

Humphrey, J. D. and Holmes, J. W., 2008. Style and ethics of communication in science and engineering. *Synthesis Lectures on Engineering*, 3(1):1–140.

Northey, M. and Jewinski, J., 2012. *Making Sense in Engineering and the Technical Sciences: Making Sense in Engineering and the Technical Sciences: A Student's Guide to Research and Writing*. OUP Canada.

Day, R. A. and Gastel, B., 2006. How to write and publish a scientific paper. Cambridge University Press.

Silvia, P. J., 2007. *How to Write a Lot: A Practical Guide to Productive Academic Writing*. American Psychological Association.

Sternberg, D., 2014. *How to Complete and Survive a Doctoral Dissertation*. St. Martin's Griffin.

Luey, B., 2002. *Handbook for Academic Authors*. Cambridge University Press.

CHAPTER 9

Safeguarding Your Personal Health and Happiness

9.1 THE CHALLENGES YOU MAY FACE IN GRADUATE SCHOOL

Life brings us different challenges at different times. Some of us live many years or even the bulk of our lives without experiencing much difficulty in our personal relationships, work, or health. For others, hardship is something that comes earlier, and potentially, more often. Regardless of your level of experience with adversity, each experience sharpens your ability to overcome obstacles and brings new opportunities for learning about yourself.

For the average person, the speed of everyday life has quickened to a frenzy. Many of us are continuously digitally connected with seemingly endless new information being thrown at us. IFLScience reports that "Ninety percent of the data in the world today has been created in the last two years alone.[1]" Exposure to the bombardment of information leads to distraction and mind wandering that can negatively impact our attention and our day-to-day fulfillment.[2] On top off all the usual stuff that the average person has to deal with, graduate school makes things a bit more amplified, with higher stress levels associated with education-related deadlines and expectations.[3]

The next few paragraphs are going to dwell on the negatives of graduate school, but I'd like to pause here and give advance notice that there are concrete steps you can take to mitigate and even eliminate these issues. In fact, if you are already well aware of the issues, please feel free to skip ahead to the next section!

Graduate study is different from undergraduate study and requires a student to make a transition in their approach to education and scholarship. Your experience will become more centered on your research activities, particularly as you progress in a Ph.D. program. It is also common for a graduate student's experience in their program to be punctuated with critical deadlines and exams that have broader career implications with limited opportunities for a "do-

---

[1]IFLScience, "How Much Data Does The World Generate Every Minute?" http://www.iflscience.com/technology/how-much-data-does-the-world-generate-every-minute/.

[2]Killingsworth, M. A. and Gilbert, D. T., 2010. A wandering mind is an unhappy mind. *Science*, 330.6006, 932–932.

[3]Hyun, J. K., Quinn, B. C., Madon T., and Lustig, S., 2006. Graduate student mental health: Needs assessment and utilization of counseling services. *Journal of College Student Development*, 47(3):247–66.

over." These exams can produce high levels of stress that have measurable biological impacts on your body.[4]

Although over 80% of Ph.D. students who are enrolled full time are funded by an assistantship in engineering disciplines, these are not high-paid positions. Additionally, part-time students may be part time simply because they can't secure an assistantship. For these reasons, financial pressures can be another source of stress for graduate students.

Because graduate school is generally attended by individuals in their prime child-bearing years, often graduate students have partners, spouses, and children. In combination with a lower income and a demanding program of scholarship, family responsibilities can be challenging to juggle. Sometimes our families and friends may not be as supportive as we would like, often this is rooted in a lack of understanding about what we are doing in graduate school and what we are trying to achieve.

The graduate student experience often has similar traits to an apprenticeship. There can be negatives that arise from having strong or even singular ties to one individual research advisor.[5] Occasionally the student's committee can help to buffer the situation, but strong committee engagement is not always present for students. Furthermore, during the course of their graduate studies, a student should undergo a transition to a junior colleague. For a variety of reason's this transition may be arrested, and the student may be trapped in a role where they have little or no say over their activities even though they have established significant expertise. As a result, having a low level of autonomy can be a big source of dissatisfaction.

9.1.1   GRADUATE STUDENT MENTAL HEALTH

Engineering Ph.D. students spend an average of 6.7 years in graduate school to complete their degree.[6] It is a long and intellectually strenuous process. Students sometimes experience "slumps" that can lead to depression. Research published in *Nature Biotechnology* reported that "…graduate students are more than six times as likely to experience depression and anxiety as compared to the general population.[7]" There has been quite a bit of research on the topic of depression in graduate school and some findings point to causes such as "..social isolation, the often abstract nature of the work and feelings of inadequacy…[8]" If you are dealing with mental health issues it is important to seek out help sooner rather than later.

---

[4]Lacey, K., Zaharia, M., Griffiths, J., Ravindran, A., Merali, Z., and Anisman, H., 2000. A prospective study of neuroendocrine and immune alterations associated with the stress of an oral academic examination among graduate students. *Psychoneuroendocrinology*, 25(4):339–56.

[5]Martin, M. M., Goodboy, A. K., and Johnson, Z. D., 2015. When professors bully graduate students: Effects on student interest, instructional dissent, and intentions to leave graduate education. *Communication Education*, 64(4):438–54.

[6]National Science Board, "Science and Engineering Indicators 2018," https://www.nsf.gov/statistics/2018/nsb20181/.

[7]Evans, B., Gastelum, B., and Weiss, V., 2018. Evidence for a mental health crisis in graduate education, *Nature Biotechnology*, 36, 282–284.

[8]Flaherty, C., 2018. Mental health crisis for grad students, *Inside Higher Education*.

Perfectionism can also be an issue that some struggle with, particularly because of the need for validation and fear of criticism that can go along with it. Perfectionism can manifest itself differently[9]: As a personal demand and expectation of oneself, as a perception that others expect perfection in you, and as an expectation that others perform to unreasonably high standards. For graduate studies, the issue of "self-oriented perfectionism" can cause a number of problems that can interfere with progress. Certainly holding yourself to high standards is good, but when those high standards require you to always portray an image of perfection to others, concealing problems and struggles that you may be having, and being unwilling to ask for help when you need it, then it is a detriment to being successful. Some graduate students suffer from the "imposter syndrome," the feeling that someone made a mistake by letting them into graduate school and at any moment they will be found out as a fraud. This can lead to the need to appear perfect to others and conceal and flaws or perceived inadequacies. A recent research study of graduate students showed that "…avoiding outward displays of imperfection was the strongest and most consistent predictor of academic problems….[10]" Whether independently or with the help of counselling, if you consider yourself a perfectionist or identify with the imposter syndrome, you need to accept the reality that everyone is imperfect. Ask for the help you need, so that you can be successful.

These previous paragraphs may sound dismal, but it is important to recognize that if you are experiencing issues you are not the only one.[11] If you find at some point in your graduate career that you are struggling—facing one or more of the above issues—it is important to seek out help. Not only will you find that other graduate students experience similar issues, but there are also people, strategies, and resources available to you, if you are willing to reach out for some help. Because universities now better recognize the issues faced by students at the undergraduate and graduate levels, there are often campus resources available. You may have a university health services that you can turn to, and you are likely to have a graduate school or dean of graduate studies office on your campus that can help you to identify the resources that are available on campus.

9.2   STEPS YOU CAN TAKE TO BE HEALTHIER AND HAPPIER

The challenging aspects of graduate school can have a detrimental effect on you as a person, but you have control over more than you may think. In particular, you have the ability to organize

---

[9]Hewitt, P. L., Flett, G. L., Sherry, S. B., Habke, M., Parkin, M., Lam, R. W., McMurtry, B., Ediger, E., Fairlie, P., and Stein, M. B., 2003. The interpersonal expression of perfection: Perfectionistic self-presentation and psychological distress. *Journal of Personality and Social Psychology*, 84(6):1303.

[10]Cowie, M. E., Nealis, L. J., Sherry, S. B., Hewitt, P. L., and Flett, G. L., 2018. Perfectionism and academic difficulties in graduate students: Testing incremental prediction and gender moderation. *Personality and Individual Differences*, 123, 223–228.

[11]Evans, T. M., Bira, L., Gastelum, J. B., Weiss, L. T., and Vanderford, N. L., 2018. Evidence for a mental health crisis in graduate education. *Nat. Biotechnol.*, 36(3):282.

your schedule and set priorities so that you have the opportunity to be both healthy and happy as you pursue your graduate studies.

For graduate students, just like other professionals, "Work—life balance is associated with physical and mental well-being…[12] The next several sections identify some of these work-life balance topics—such as getting exercise and eating healthy—and the strategies that can help you achieve your goals—such as mindfulness practice and time management.

Some of what you need to do is simple. In Claire Potter's essay outlining "The Ten Commandments of Graduate School[13]" her second commandment, after "Thou shalt no rack up unnecessary credit card debt," advises that you not neglect your dental and health care. If you move to a new location for graduate school, you will need to set up new doctors and dentists for yourself. Determine what your insurance benefits are—at some institutions you will have coverage—and find out who the providers are. Don't wait until you have a crisis, get established with a new doctor and a new dentist early. Additionally, there are numerous campus resources that can help you to navigate specific health issues that you are already aware of, or may arise in the future.

Part of your baseline for happiness in graduate school is the people who you interact with on a day-to-day basis. Recall back to Chapter 2 on *Finding the Right Research Position for You*. It has been shown that having a "…strong, supportive and positive mentoring relationships between graduate students and their PI/advisors correlate significantly with less anxiety and depression.[14]" Knowing that is helpful for making a good choice at the start, but even if you find that you don't have the kind of relationship you would have wished for with your research mentor, it is not the end of hope. Focus on broadening your constellation of mentors to find the support you need to succeed. On occasion, however, some graduate students find themselves in a position that is negative and destructive, and a change of research mentor is needed. If the relationship is one that you need to remove yourself from, it does not mean that you have to give up your goals for achieving your Ph.D. Work with trusted colleagues on your campus to help you identify a better path forward (for instance, many campuses have an Ombudsperson who you can consult with confidentially).

> **Changing Course**
> I have had at least one case in my research group where in the process of mentoring we discovered that the alignment between the student's newly refined goals and the research that was being conducted in my lab were not as good as we once thought they were. In this case, I helped the student to identify the new direction that they would like to take and assisted them in

---

[12]Evans, B., Gastelum, B., and Weiss, V., 2018. Evidence for a mental health crisis in graduate education, *Nature Biotechnology*, 36, 282–284.

[13]Potter, C., 2013. The ten commandments of graduate school, *Chronicle of Higher Education*.

[14]Flaherty, C., 2018. Mental health crisis for grad students, *Inside Higher Education*.

getting to where they wanted to be. Although this is a problem in the short term for me, and a loss because I have invested both time and funding into their training, I have found that in the long run it's best for both the people and the project to make the change.

I know that changing research groups was a difficult conversation for my student to initiate with me. Part of what helped to make it work was their willingness to help us complete our short-term goals on the research project while we were looking for a better long-term path for the student.

ASSIGNMENT 9-1:
INDIVIDUAL ASSIGNMENT – IDENTIFYING SUPPORT RESOURCES

Nearly every graduate school in the U.S. has support resources that their graduate students can take advantage of.[15]

This may include access to workshops on stress management, child care sharing groups, individual mental health counselling sessions, support groups or boot camps on dissertation writing, athletic facilities, non-credit classes offered by the union or continuing studies, just to name a few. Investigate the resources available to you on your campus (use websites, graduate student coordinators, and fellow students). Identify at least two resources that you would find personally beneficial immediately, and two additional resources that you could envision benefiting from in the future when you are at a different stage in your graduate career or experiencing a specific struggle. Choose one of these resources that you can utilize this week and make an appointment and/or schedule time for it in your calendar.

9.3   GETTING SLEEP

It's easy to let the end of the term, a looming deadline, or an important degree program exam, get your schedule out of whack. But that's actually the worst time to get less sleep. Having a consistent sleep routine and sleep schedule are important for both your physical and mental health.

Lacey et al. found that "…during the course of lengthy anticipatory periods preceding a scheduled oral examination, graduate students reported more frequent malaise (e.g., headaches, sore throat, fatigue) than did controls." Furthermore, "…anticipation of an imminent oral aca-

---

[15]Bird, L. T. and Sheryl, A. Principles of Good Practice in Dealing with Students in Distress: Council of Graduate Schools. Available from: http://cgsnet.org/principles-good-practice-dealing-students-distress-0.

demic examination was also associated with increased cortisol levels[16]"—a hormone that regulates important bodily functions like metabolism and immune system response. Even pulling one all-nighter or getting minimal sleep before an exam can be detrimental. Alterations of immune function occur after only a modest loss of sleep.[17]

Good "sleep hygiene" begins with taking care of your body during the day and allowing your brain to cool down before you turn off the light to go to sleep. You have likely heard much of this advice before, but it is important to avoid caffeine in the later part of the day and alcohol before bedtime. You also need to put away the screens and do a relaxing activity like meditation, journaling, or reading (from paper) before you turn off the lights. Make sure your sleeping situation is comfortable, dark, and quiet (if not, an eye mask and earplugs can be helpful). Once you have developed a routine stick with it. Some people have no trouble going to sleep at the beginning of the night but can't get a full night's sleep because they wake before they intend to and have difficulty going back to sleep. If this happens to you and your mind is racing, it may be helpful to keep a notepad by your bedside to write down what you are thinking about so you can let go of it for now and get back to sleep. You might also find that re-engaging with aspects of your bedtime routine, like meditation of reading, may help you to return to restful sleep.

ASSIGNMENT 9-2:
INDIVIDUAL ASSIGNMENT – PERSONAL SLEEP LOG

You may not realize how irregular your sleep pattern is if you are not attuned to the issue. Place a paper calendar and pencil next your bedside. Each morning jot down the approximate time you fell asleep the previous night, the time at which you woke up, the total hours of sleep, and a quality rating of your sleep between 1 and 10. Do this during a representative week of the semester. For the next week identify a target time at which you will go to bed every night and try to maintain that nighttime routine while continuing to record data. At the end of the second week identify whether your sleep quality improved. Use subsequent weeks to experiment with other sleep improvement techniques, such as limiting exposure to TV/computer/phone screens before bedtime, making modifications to your sleep environment to ensure that it is dark and quiet, and avoiding caffeine during the second half of the day.[18]

[16]Lacey, K., Zaharia, M., Griffiths, J., Ravindran, A., Merali, Z., and Anisman, H., 2000. A prospective study of neuroendocrine and immune alterations associated with the stress of an oral academic examination among graduate students. *Psychoneuroendocrinology*, 25(4):339–56.

[17]Irwin, M., Mcclintick, J., Costlow, C., Fortner, M., White, J., and Gillin, J. C., 1996. Partial night sleep deprivation reduces natural killer and cellular immune responses in humans. FASEB 10, 643–653.

[18]Mayo Clinic, "Sleep tips: 6 steps to better sleep," https://www.mayoclinic.org/healthy-lifestyle/adult-health/in-depth/sleep/art-20048379.

9.4   GETTING EXERCISE

It's important to get exercise, particularly if your coursework and research keep you pinned at a desk most of the day. In addition to being good for your body, exercise can help you reduce stress and can be beneficial for your brain. Depending on what you decide to do for exercise, it also the potential to help you meet new people beyond your research group and graduate program and develop new friendships. These can be a valuable support network for you.

If you have moved to a new location for your graduate studies, the types of exercise you used to rely may not be as readily available because of a different climate or access to facilities. Take the opportunity to expand your horizons—try out new sports, identify clubs, and test different activities. Many campuses have club sports, leagues, and even non-credit classes that can help you to test out something new and develop skills. Minimally, most campuses have some sort of gym/pool access available to students.

Think broadly about what might be available locally: running and/or biking trails, hiking trails, ice skating, cross-country ski trails, downhill skiing, sailing, rowing, kayaking, rock climbing, swimming, dancing, etc. Consider club and sport teams like baseball/softball, volleyball, lacrosse, kickball, and even quidditch.

For some people exercise is already a part of their daily routine, but for others it is something we have to push ourselves to do regularly. If you are in the second category, there are a number of strategies that might work for you. Try scheduling exercise into your calendar just like you would do for a course, sign up for a non-credit class that meets regularly, or find an "exercise buddy" who will help you get out to exercise regularly.

If you're starting a new exercise routine or sport, start out realistically and slowly ramp up to a level that is healthy and sustainable. If you have concerns about how exercise may impact past injury or other health condition, consult with your physician before embarking on anything strenuous.

ASSIGNMENT 9-3:
INDIVIDUAL ASSIGNMENT – CAMPUS SPORT AND RECREATION RESOURCES

Access your institution's website and determine what campus resources are available to you for getting/staying fit. Search on terms like "recreational sports" and "club teams." Identify several that would be of interest to you and choose one to check out in person. Get a facility tour or meet with someone who will provide you with an orientation.

9.5   EATING HEALTHY

One of the challenges of being a student, particularly if you live on or near campus, is finding ways to consistently eat healthy. Pizza deliveries, fast food restaurants, and sandwich shops are readily available, but these do not generally provide the kind of food that will help you stay healthy and fuel your brain effectively. Unfortunately, many campuses meet the definition of a food desert, thus it is difficult to easily obtain healthy, fresh food. In response some campuses have welcomed area farmers to hold small farmer's markets on campus during the growing season, this can be a great way to insert more fresh produce into your diet.

In some areas of the country, community-supported agriculture (CSA) provides a way for you to both support a local farm by buying directly from them and receive a box of fresh produce weekly during the growing season. Depending on the particular CSA you join, you may be able to choose the size of box, delivery frequency, and even the choice of items. Some farms have drop off locations on university campuses.

There may also be opportunities to buy fresh ingredients for healthier eating just a bus ride away. Look into the areas grocery stores that are available and the public transportation options that are connected to campus.

If you are on a land-grant campus or a university with a large agricultural program you may also have the opportunity to buy food from campus sources. At the University of Wisconsin-Madison for instance, the Meat Sciences Laboratory has Bucky's Butchery shop (an undergraduate operated store that sells meat one day a week), the Babcock Dairy Store has award-winning cheese (and phenomenal ice cream), and the UW Poultry Science Club sells turkeys each Thanksgiving. Take a look into what your campus has to offer, you might be surprised at what you find.

You may have to do a bit of internet sleuthing, but there are likely some good options to help you eat healthy that are more easily accessible than you might have initially appreciated!

ASSIGNMENT 9-4:
INDIVIDUAL ASSIGNMENT – HEALTHY FOOD EXPLORATION

Identify a fresh vegetable available to you locally that you have never tried before or don't normally eat. Use this vegetable as a search term in your favorite cookbook or in an online recipe resource (e.g., www.allrecipes.com). Find a recipe that looks appealing, buy the ingredients, and give it a try.

9.6   CREATIVE OUTLETS

The creative nature of engineers enables their ability to innovate and discover. To imagine what might be possible.

Engineers often express their creativity in a number of ways outside of their engineering practice. If you ask your engineering colleagues you may find that you are among musicians, dancers, writers, painters, potters, woodworkers, and more. If you are one of these engineer artists, be sure to allow time for your creativity both inside and outside of engineering. Not only is it a good stress relief, you may find that is a helpful way to practice achieving a state of immersion that you also need to be productive with your engineering data analysis, technical writing, etc.

Your arts practice may also ultimately link with your engineering work in ways you may not have initially anticipated.[19] The more you focus on observing, the more you will see. Your thinking and skills and creativity will be enhanced not only by improving your math skills and your language skills, but also your perceptual skills.[20]

For example, engagement with visual representation has been crucial aspect of my professional career. My training as a visual artist has been essential skill building that has helped me to understand and interpret the images obtained from a variety of microscopy instrumentation. The ability to see detail and attend to subtle changes in images is critical to my engineering research. I believe that these skills are enhanced by my artistic practice with painting. In my teaching, I use visual representations of engineering elements and concepts. They are integrated throughout the courses that I teach at the undergraduate and graduate levels to provide learners with additional ways of interacting with complex concepts.

The Pause that Refreshes

I have always enjoyed art as a hobby and had taken painting and sculpture classes prior to attending graduate school. During my Ph.D. program I enrolled in a ceramics class offered through the Art Department (it was pretty intense, so I chose the pass/fail option for the course even though I probably could have earned a good grade). After taking the class I realized that being able to immerse myself in art periodically was reducing my stress overall and helping me to be more focused when I came back to my engineering studies and research. I discovered that the campus student union also had non-credit classes and you could have access to the studio, pottery wheels, and kilns by paying a small fee even if you were not enrolled in a class. In the studio I bumped into a fellow graduate student studying chemistry who enjoyed ceramics as well. She and I began meeting regularly at the studio to work on our pottery. It was a wonderful complement to my engineering work that I continued throughout my Ph.D. program.

---

[19]Walesh, S. G., 2019. Can creating art make you a more effective engineer?, *PE Magazine*, National Society of Professional Engineers, pp. 24–27.

[20]Edwards, B., 2008. *Drawing on the Artist Within*, Simon and Schuster.

9.7   EMPLOYING MINDFULNESS PRACTICES

There is ample scientific research that the use of regular mindfulness practices, such as meditation, can have a positive impact on our body and make changes in how our mind works.[21]

What are mindfulness practices? "An operational working definition of mindfulness is: the awareness that emerges through paying attention on purpose, in the present moment, and nonjudgmentally to the unfolding of experience moment by moment.[22]" One way to achieve the qualities of attention and awareness, thought of as being characteristic of mindfulness, is the practice of meditation. Historically, meditation has been connected to Hinduism and Buddhism, but more recently it has been Westernized and converted into a secular practice.

Mindfulness practices involve two basic components: "The first component involves the self-regulation of attention so that it is maintained on immediate experience, thereby allowing for increased recognition of mental events in the present moment. The second component involves adopting a particular orientation toward one's experiences in the present moment, an orientation that is characterized by curiosity, openness, and acceptance.[23]" Mindfulness practices are broader than just meditation. Other mindfulness practices you might be familiar with include yoga and Tai Chi.

There is mounting scientific evidence that regular mindfulness practice such as meditation can change your brain and your body. Studies that ask participants to employ daily meditation show that individuals can manage chronic pain, reduce stress hormones, and improve their resilience. There is a growing literature showing that activities like Tai Chi, Qigong, yoga, and meditation can alter inflammatory gene expression and change cellular markers of inflammation, even after just 6–8 weeks of training and practice.[24]

You should not feel that a major lifestyle change is required to achieve some benefit. Small amounts of regular meditation can also be helpful to reduce stress and improve your capacity for creative thinking.[25] One simple mindfulness practice is a focus on the breath. For example:

> Sitting in a comfortable position, you close your eyes and notice your breath. It is sometimes easier to focus by using a count with your breathing. Breath in counting to one, and breath out to one, breath in counting to two and breath out to two, breath in counting to three and breath out three, and so on. There will be a point where you find the count length to be uncomfortable, so then reverse your count until you

[21]Davidson, R. J. and Begley, S., 2012. *The Emotional Life of Your Brain: How Its Unique Patterns Affect the Way, Your Think, Feel, and Live—and How You Can Change Them*, Plume, New York.

[22]Kabat-Zinn, J., 2003. Mindfulness-based interventions in context: past, present, and future. *Clinical Psychology: Science and Practice*, 10(2), 144–156.

[23]Bishop, S. R., Lau, M., Shapiro, S., Carlson, L., Anderson, N. D., Carmody, J., Segal, Z. V., et al., 2004. Mindfulness: A proposed operational definition. *Clinical Psychology: Science and Practice*, 11(3):230–241.

[24]For example: Ader, R., Cohen, N., and Felten, D. L., 1987. Brain, behavior, and immunity. *Brain, Behavior, and Immunity* 1(1):1–6. And Bower, J. E. and Irwin, M. R., 2016. Mind—body therapies and control of inflammatory biology: a descriptive review. *Brain, Behavior, and Immunity*, 51, 1–11.

[25]Schootstra, E., Deichmann, D., and Dolgova, E., 2017. Can ten minutes of mediation make your more creative? *Harvard Business Review*.

find a comfortable count length for your breath and continue breathing in and out at this comfortable rate. As you continue to breath in and out, things will pop into your mind and that's ok. Just take note of it, let the idea pass without dwelling on it, and then refocusing your mind on your breath. When you feel you are ready to stop, simply open your eyes.

This sort of practice allows you to enhance your focus, select what you choose to focus on in the moment, and build the ability to notice your thoughts objectively without further elaboration.

There are a wide range of skills and techniques that you can use to build mindfulness into your everyday life. I highly recommend that students consider getting training in one or more mindfulness techniques. There are often opportunities available on or near university campuses, as well as a plethora of online resources[26] and aps that you can use.[27]

ASSIGNMENT 9-5:
INDIVIDUAL ASSIGNMENT – MINDFUL RESET, MINDFUL RECHARGE, MINDFUL REFRESH

Use brief mindfulness activities during your day to reduce your stress and increase your productivity.

Make a copy of the activity chart (Table 9.1). Cut along the lines and place the pieces in an envelope. The next time you are feeling stuck, or stressed, or just need a break, pick an activity card at random from the envelope and spend some time practicing your mindfulness skills with the activity described.

9.8   MAKING TIME FOR IT ALL

In Chapter 5 we discussed strategies for project management as it relates to your research. For some of us it might be helpful to employ project management tools with aspects of your personal life too.

---

[26] For example:
• Center for Healthy Minds, University of Wisconsin–Madison (resources on cultivating wellbeing and relieving suffering through a scientific understanding of the mind, including some guided practices):
http://centerhealthyminds.org/join-the-movement/workplace
• Center for Advanced Studies in Business, University of Wisconsin–Madison (guided audio practices):
http://www.uwcultivatingwellbeing.com/guided-audio-practices
Inner Sense Consulting, Bev Hays (guided mindfulness meditations):
http://www.innersenseconsulting.com/meditations.html.
[27] Two of my current favorites:
• Simply Being
• Stop, Breathe and Think

Table 9.1: Mindful Reset, Mindful Recharge, Mindful Refresh Activities (*Continues.*)

Meditate	Sense	Draw
Meditate for 10 minutes on the present moment. Find a comfortable position, sitting upright if possible, and bring yourself into stillness. Pay close attention to your breath. Note, but don't dwell on, the thoughts, emotions, and sensations that occur.	Translate music into drawing. Listen to a piece of music that you enjoy. Translate the feeling that music gives you into a drawing – abstract or realism.	Find an object in your environment that is familiar to you. While looking at the object, do a line drawing of the object while NOT looking at the paper. Imagine that your pencil or pen is actually touching the contour of the object as you draw.
Journal	**Exercise**	**Meditate**
Think about ways in which you are committed to the common good. Choose one of these and write about what you will do in the future to advance that common good.	Walk 5+ flights of stairs. Count each step. (Pay attention to your body and stop if you feel pain.)	Close your eyes, take slow deep breaths, and count 20 of them. If your mind wanders, take note of what you were thinking about, and then bring yourself gently back to focus on your breath.
Journal	**Connect**	**Journal**
Write about 3 things that give you a genuine smile.	Spend some time talking to someone you have not spoken to in a while.	Write down 20 things that you are grateful for.
Draw	**Journal**	**Sense**
Fill an entire piece of paper with repetitions of a shape or pattern.	What was the best thing that happened to you today? Take a few minutes to write about it.	Eat a healthy snack while giving the experience your full attention. Focus on the taste, texture, and aroma.

Table 9.1: (*Continued.*) Mindful Reset, Mindful Recharge, Mindful Refresh Activities

Connect	Sense	Journal
Send a brief note/message of appreciation to a colleague, friend, or family member.	Take a walk in nature. Find a small space or a large expanse where you can observe the flora and fauna around you.	When did you feel the most proud of yourself today? Take a few minutes to write about it.
Meditate	**Compose**	**Exercise**
Take a mindful walk in a safe place. Walk a bit slower than your usual pace. Focus your awareness on your movements, balance, and the rhythm of your steps. If your mind wanders, take note of what you were thinking about, and then bring yourself gently back to focusing on your walk.	Create a theme song for your day. Consider which word or topics repeat as you think about the positive ways in which you would like your day to progress. Use these words or topics to create a chorus for your theme song.	Do sitting isometric exercises. For instance, while sitting and keeping your knees bent at a right angle, pick up one foot off the floor for a count of 10. Switch and lift the other foot for a count of 10. Pay attention to your body during each exercise and stop if you feel pain.
Compose	**Journal**	**Sense**
Translate feeling into rhythm. Compose a short rhythm expressing a feeling you have chosen. Express it with finger snapping, toe tapping, tongue clicking, etc.	Think of a time when you were resilient in the face of adversity, small or large. Focus on that resiliency and write about how you can use this resiliency in the future.	Find a pleasant scent in your environment. Close your eyes and inhale deeply. Concentrate on the sensations and thoughts that spring to mind.
Sense	**Draw**	**Exercise**
Find something that feels cool or warm to the touch. Place your hands on the object. Close your eyes and engage in the moment.	Do a sketch of something in your field of view. Focus on the energy or mood of what you are drawing.	Do yoga stretches for 5 minutes. Three poses. (Pay attention to your body and stop if you feel pain.)

Different people have different challenges when it comes to making time for it all. For some of us we have too little "me time" and for others there is too much "play time." All of us need a balance. You need to devote substantial time to making progress in your graduate education, and you will be able to more effectively do so, if you have a healthy mind and body. It is also important to resist making comparisons between yourself and other students in your degree program. Each person's path is a unique one. Your goal is to find the most effective and efficient path for yourself, that allows you to achieve your goals while also being a well-balanced individual over time.

ASSIGNMENT 9-6:
INDIVIDUAL ASSIGNMENT – PLANNING YOUR WEEK

In reality, every week of your life will be different from the next and you will have to be flexible as a deadline or special event approaches. However, you can develop some principles that you would like to follow in managing your time on a regular basis. Begin by creating a list of the major activities that you undertake regularly, e.g., coursework, if you are taking classes, research activities, including writing, personal time that includes exercise, and other activities that make you either relaxed or energized, social/family time, and sleep. Now set out goals for the number of hours a week you feel you should devote to each activity. Schedule how this should look on a weekly calendar for the "average" week (Table 9.2).

Now plan your actual calendar for the coming week. Live your life, do your work, and at the end of each day jot down the number of hours you spent on each activity category. At the end of the week total up how you spent your time and compare it to your goals. You may find that you have not set your goals realistically, or you may find that this week had unexpected events. Respond to this by balancing out how you spend your time in the following week. Continue this planning activity for four weeks.

This process is designed to help you find a routine that includes all of the activities you should spend time on, and a sufficient amount of time on them to make progress and meet your deadlines. Be sure you have not eliminated sleep, personal, and social/family time to make time for everything else—without balance you will be less productive overall.

Table 9.2: Weekly calendar used to set goals for the number of hours planned for each activity and actual time spent on each activity daily

Planned / Actual	Research	Course/ Teaching	Communication/ Networking/ Service	Relax/ Exercise/ Recharge Activities	Family/ Personal Responsibilities	Sleep
Monday						
Tuesday						
Wednesday						
Thursday						
Friday						
Saturday						
Sunday						
Total						

Afterword

This book is based on my experiences as a research mentor, graduate advisor, instructor, and administrator in the Graduate School of the University of Wisconsin–Madison. I am grateful to all of the undergraduate and graduate research assistants who have worked with me over the years, not only for their research contributions, but also for how they helped me to develop and learn as a mentor. My teaching in the Engineering Physics undergraduate program and the phenomenal undergraduate honors students I have worked with have helped me to better understand what novice researchers want and need to know as they begin their research career. I also had the pleasure of serving in several different administrative roles in the University of Wisconsin–Madison Graduate School for five years, where I provided leadership for all aspects of the graduate student experience, including admissions, academic services, academic analysis, funding, professional development, and diversity. I learned an immense amount from my colleagues in the Graduate School and my faculty and staff colleagues across the University who devote time and energy to graduate education. Those experiences and interactions also allowed me to see graduate education from a broader perspective, beyond that of the graduate programs in the College of Engineering where I serve as a graduate advisor and research mentor. This book draws from this range of experiences and offers guidance and advice to those entering engineering research as an undergraduate or a new graduate student.

Author's Biography

WENDY C. CRONE

Wendy C. Crone is the Karen Thompson Medhi Professor in the Department of Engineering Physics, with affiliate faculty appointments in the Department of Biomedical Engineering and the Department of Materials Science and Engineering, and she holds the honor of Discovery Fellow with the Wisconsin Institute for Discovery at the University of Wisconsin–Madison.

Her research is in the area of solid mechanics, and many of the topics she has investigated are connected with nanotechnology and biotechnology. She has applied her technical expertise to improving fundamental understanding of mechanical response of materials, enhancing material behavior through surface modification and nanostructuring, exploring the interplay between cells and the mechanics of their surroundings, and developing new material applications and medical devices. Her research has been funded by the National Institutes of Health, National Science Foundation, Department of Energy, Air Force Office of Scientific Research, and Whitaker Foundation.

She teaches courses in the areas of engineering mechanics, engineering physics, and informal science education. Over the last two decades, Prof. Crone has trained over two dozen graduate students and postdocs in engineering mechanics, materials science, biomedical engineering, and engineering education. Her former students hold positions in academia, national laboratories, and industry.

Prof. Crone has received awards for research, teaching, and mentoring. In addition to numerous peer reviewed journal publications, dozens of explanatory education products, and four patents, she is the author of the book *Survive and Thrive: A Guide for Untenured Faculty*. She has also served in several leadership roles over the course of her career, including Interim Dean and Associate Dean of the Graduate School at UW-Madison and President of the Society for Experimental Mechanics.

Index

Printed in the United States
by Baker & Taylor Publisher Services